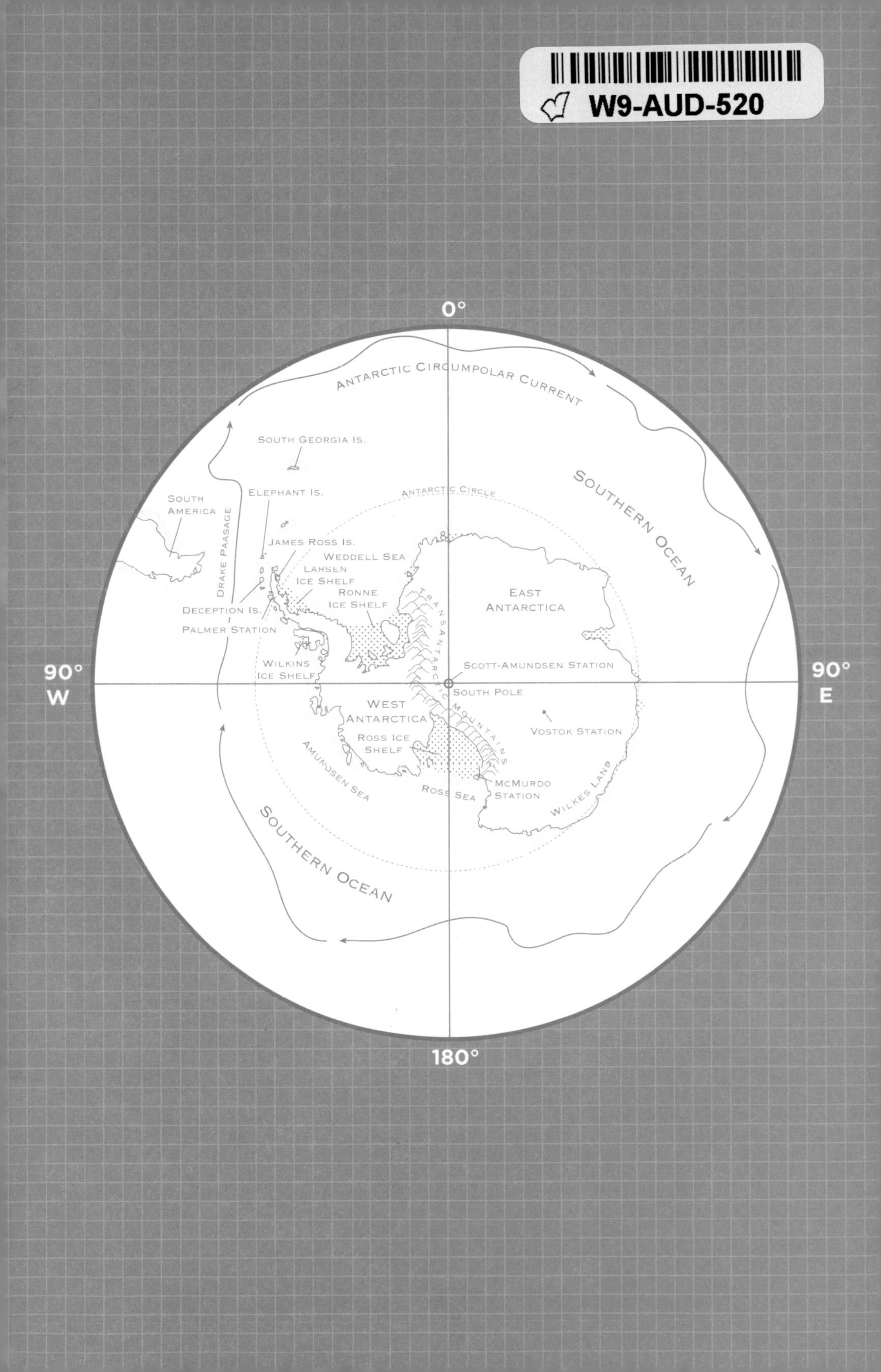
W9-AUD-520
0°
Antarctic Circumpolar Current
South Georgia Is.
Southern Ocean
South America
Elephant Is.
Antarctic Circle
Drake Paasage
James Ross Is.
Weddell Sea
Larsen Ice Shelf
Ronne Ice Shelf
East Antarctica
Transantarctic Mountains
Deception Is.
Palmer Station
Wilkins Ice Shelf
Scott-Amundsen Station
90° W
90° E
South Pole
West Antarctica
Vostok Station
Ross Ice Shelf
Amundsen Sea
Ross Sea
McMurdo Station
Wilkes Land
Southern Ocean
180°

A WORLD
WITHOUT
ICE

HENRY POLLACK, PH.D.

AVERY | *a member of Penguin Group (USA) Inc.* | *New York*

Published by the Penguin Group
Penguin Group (USA) Inc., 375 Hudson Street, New York,
New York 10014, USA • Penguin Group (Canada), 90 Eglinton
Avenue East, Suite 700, Toronto, Ontario M4P 2Y3, Canada (a division
of Pearson Penguin Canada Inc.) • Penguin Books Ltd, 80 Strand,
London WC2R 0RL, England • Penguin Ireland, 25 St Stephen's
Green, Dublin 2, Ireland (a division of Penguin Books Ltd) • Penguin
Group (Australia), 250 Camberwell Road, Camberwell, Victoria 3124,
Australia (a division of Pearson Australia Group Pty Ltd) • Penguin
Books India Pvt Ltd, 11 Community Centre, Panchsheel Park,
New Delhi–110 017, India • Penguin Group (NZ), 67 Apollo Drive,
Rosedale, North Shore 0632, New Zealand (a division of Pearson
New Zealand Ltd) • Penguin Books (South Africa) (Pty) Ltd,
24 Sturdee Avenue, Rosebank, Johannesburg 2196, South Africa

Penguin Books Ltd, Registered Offices:
80 Strand, London WC2R 0RL, England

Published simultaneously in Canada

Library of Congress Cataloging-in-Publication Data

Pollack, H. N.
A world without ice / Henry Pollack, Ph.D.
p. cm.
Includes bibliographical references and index.
ISBN 978-1-58333-357-0
1. Glaciers. 2. Ice 3. Global warming. I. Title.
GB2405.P55 2010 2009030326
551.31'2—dc22

This book is printed on recycled paper.

Printed in the United States of America
1 3 5 7 9 10 8 6 4 2

BOOK DESIGN BY NICOLE LAROCHE
ILLUSTRATIONS BY KIM CHATER
MAPS BY DALE AUSTIN

To my students and colleagues
who have helped me understand how Earth works

CONTENTS

FOREWORD

Humanity has arrived at a historic moment of decision. Our home, Earth, is in grave danger. At risk of destruction is not our planet itself, but the unique climatic conditions that have made it hospitable for human life and that help sustain civilization as we know it.

For decades now, scientists have studied Earth's climate, the causes contributing to its warming, and how this rising fever is impacting and will continue to impact our world and way of life. As overwhelming scientific evidence mounted that human activity is the primary cause of dramatic and accelerating global warming, many in the scientific community raised the alarm.

Testifying before elected officials and policymakers, collaborating on international studies, reporting to the United Nations Intergovernmental Panel on Climate Change, consulting with business leaders and speaking out in the media, scientists around the world have worked hard to communicate the severity and immediacy of the challenge before us—a challenge of our own inadvertent making and one requiring urgent, bold, and united action to address.

The consequences of inaction are already becoming more and more apparent: rising sea levels, more severe droughts, increasingly violent storms, the spread of disease, the loss of crops, disappearing wildlife, and politically destabilizing tides of climate refugees.

Still, a small group of climate change naysayers—some with deep financial interests in defending the status quo, and others with philosophical objections to any role for government in solving the crisis—have mounted a vigorous public relations campaign: first, to try to refute the inconvenient truth of global warming; second, to question its causes; third, to minimize its consequences; and finally, to complain about the alleged costs of transitioning from fossil fuels to more sustainable and renewable energy sources.

Such dissembling and obstruction have taken their toll; over the past fifteen years, as the United States refused to play a leadership role in marshaling global efforts to combat climate change, the crisis has grown much more severe. Now we approach several global tipping points, which—if we do not address them boldly and immediately—may plague generations to come for literally thousands of years.

One such tipping point is the loss of Earth's ice, accumulated over millions of years but now melting at an alarming rate. Across the globe, glaciers that have for centuries provided agricultural and drinking water for more than a billion people are quickly disappearing. In Antarctica, ancient ice shelves—some the size of Belgium, Scotland, or France—are beginning to disintegrate, sending massive icebergs into the Southern Ocean. Hundreds of glaciers on Greenland are slipping faster and faster into the Atlantic. And in the Arctic Ocean, where an ice sheet has capped the polar sea for not only the entire history of human civilization but for most of the last three million years, researchers have in recent years observed a fast-paced shrinking and thinning of the Arctic sea ice that will in just a decade lead to an ice-free Arctic Ocean throughout the summer months.

Some might ask—so what? Why should we care about vanishing glaciers and remote polar ice caps? Why does the disappearance of distant ice really matter to a farmer in Nebraska, an engineer in Rio de Janeiro, an investment banker in New York, a salesperson in Cincinnati, a bus driver in London, a tour guide in Tanzania, a child in Bangladesh, or a businessman in Beijing?

The answer is that, on a global scale, ice plays a critical and major role in setting the temperature of Earth's atmosphere and oceans, governing major weather patterns, regulating sea level, and dramatically impacting agriculture, transportation, commerce, and even geopolitics. Though we have been slow to recognize this, our history and future are inextricably linked to the world's ice.

If we do not act now, as individuals, as communities, as businesses, and as nations to slow and gradually halt the current meltdown, we risk destroying the very global systems that have enabled us to thrive and prosper.

In this insightful, accessible, and important new book, Dr. Henry Pollack explores the vital role that ice plays in the functioning of our planet, how it impacts human life, how we in turn are affecting ice, and why the decisions we make today, both individually and collectively, will shape the world and human society for thousands of years to come.

A World Without Ice explains complex global systems in simple terms without dumbing down the message, and explores the very real implications for people and the planet without succumbing to romanticism or hyperbole. As such, this book will help a broad range of readers understand the challenges we face and the high stakes of the climate change debate. Readers will be confronted with the fact that humans are no longer passive players in the global climate system, but rather key drivers of many changes now becoming apparent.

I have known Dr. Pollack and followed his research into global climate change since 1992, when I called on him to testify before the Senate, and later I sought his expertise at the White House. A geophysicist at the University of Michigan since 1960, Dr. Pollack led a global scientific consortium that reconstructed Earth's surface temperature over the past millennium, through the innovative use of borehole temperature measurements made in the rocks of Earth's crust.

He has advised the National Science Foundation, testified before

National Academy of Sciences and congressional committees, and was a contributor to and member of the United Nations Intergovernmental Panel on Climate Change, which shared the 2007 Nobel Peace Prize.

In addition to respecting his scientific expertise, though, I have also come to know and respect Dr. Pollack as an effective communicator, a scientist with the rare ability to engage ordinary people and to translate scientific ideas into everyday terms that are easy to understand. I've watched Dr. Pollack, as one of my advisors on The Climate Project, help hundreds of people wrap their minds around the complexities and challenges of climate change, and come away not only better informed, but feeling empowered by their knowledge as well. Intuitively, he grasps the relationship between science and humanism, and bridges them with rationality and grace. He brings that same straightforward, affable approach to the pages of *A World Without Ice*, a book that I believe will help large numbers of readers understand and meet the paramount challenge of our time.

For years now, efforts to address the growing climate crisis have been undermined by the idea that we must choose between our planet and our way of life, between our moral duty and our economic well-being. These are false choices. In fact, the solutions to the climate crisis are the very same solutions essential to redressing our economic and national security crises.

Though daunted by the crisis at hand, I am by nature an optimist. I have seen people's capacity, when informed and inspired, to bring about necessary change. I am hopeful that *A World Without Ice*, in illuminating the challenge of climate change through the prism of ice, will help spark people to engage their friends and families, their communities, congregations, companies, and countries, in the battle to save the natural environment that has nurtured us. We live on an unlikely, infinitesimal island of possibility in the endless sea of space, and we must all work to save our home before it's too late.

—AL GORE

February 11, 2009

PREFACE

This is a book about ice and people on Earth—the impact ice has had on our planet, its climate, and its human residents, and the reciprocal impact that people are now having on ice and the climate of the future. Ice has been on Earth much longer than people have—we are relative newcomers to the terrestrial menagerie. Humans have called Earth home for only some three million years, whereas ice has been a part of Earth's landscape for billions of years.

Throughout most of Earth's history, ice has been an indomitable force of nature. The creep of ice over the continents during past glacial epochs has profoundly shaped Earth's surface. The sharp Alpine peaks of Europe, the vast Great Lakes of North America, the majestic valleys of California's Yosemite National Park, and the deeply incised fjords of Norway—all are products of earlier glacial erosion. Today they grace Earth's landscape as gifts from ice to humanity. Minute by comparison, people stand awestruck at the immense scale of nature's handiwork.

But ice is much more than just a landscape sculptor and earth mover—it is a major player in Earth's climate system. Of the sunshine falling onto Earth, about 30 percent of it is reflected back into space, mostly by white clouds in the atmosphere and white ice at the surface.

The polar ice caps, covering virtually all of Antarctica, the Arctic Ocean, and Greenland, make up less than one tenth of Earth's surface, but account for much of the sunshine reflected from the surface. Polar ice also generates huge wind streams that spill ferociously off the ice caps and flow far beyond the ice perimeter to shape weather systems that influence the entire globe.

Geologists have translated the book of rocks, layer by layer, and discovered that there have been many times in Earth's history when the climate was different from that of today—episodes when ice blanketed half the globe, and times when the polar regions were free of ice. It is almost incredible to think that just twenty thousand years ago, at the places where New York, Detroit, and Chicago are today home to millions of people, the landscape was monochrome white, a massive sheet of ice a half mile thick. And there were no people in all of North America to see it, to marvel at it, or to cope with it. The Western Hemisphere had yet to welcome its first human immigrants.

Over the span of only the last three centuries, however, the rapid growth of the human population and the rise of industrial society have brought the relationship between ice and humankind to a precarious tipping point. Gone are the days when ice was unfazed by the few people living on its fringes. Today human activities are having a profound effect on Earth's climate and destabilizing the world's ice. Climate scientists warn that in the not-too-distant future we may see a world without ice.

It is difficult to envision a world without ice—it requires a stretch of the imagination no less than envisioning a world without trees, or flowers, or animals. The loss of ice will have dramatic consequences for planet and people alike. The drinking water and agricultural water for almost one quarter of Earth's population—a number that exceeds the population of the entire Western Hemisphere—come directly from mountain glaciers. An even greater number of people depend on the seasonal replenishment of water from the melting of winter snow to nourish crops at the outset of the growing season. The dramatic shrinking of Arctic sea ice over the past few decades has already triggered international posturing over oil and minerals that perhaps will be discovered on the ocean floor. And the likely disappearance of summer sea ice later in this century will set the stage for exploitation of the Arctic's fisheries, and will open maritime trade routes such as the fabled Northwest Passage between Europe and East Asia.

In the starkest terms, however, the melting of the ice now on the continents means adding more water to the oceans, and rising sea levels. The ensuing flooding will affect low-lying regions of all nations with a seacoast—more than a hundred countries. The loss of property and agricultural land, the damage to coastal infrastructure, and the pollution of fresh groundwater aquifers with salty seawater are all significant consequences with grave economic implications. But the most severe consequence will be the displacement of many millions of people who live near the sea. A sea level rise of only three feet would transform more than one hundred million coastal residents into climate refugees. Such a population displacement, a number equivalent to a third of the population of the United States, would be unprecedented in human history.

What can be done to forestall such consequences? One must recognize that some changes accompanying a warming world are unavoidable, because of the earlier inadvertent changes to the climate system that have already occurred and will continue to play out throughout this century. And it is equally clear that if no mitigation measures are taken, a broad spectrum of serious consequences will appear sooner, many within this century, and will grow larger with time. But there are a number of middle paths, some timid, some bold, that are being readied for the nations and peoples of the world to consider and perhaps embrace. The creativity of people has the potential to slow and even reverse the changes in global climate now taking place. However, the willingness of nations to take the difficult but necessary steps to do so has yet to be convincingly demonstrated.

CHAPTER 1
DISCOVERING ICE

The ice was here, the ice was there,
The ice was all around;
It cracked and growled, and roared and howled,
Like noises in a swound!

—SAMUEL TAYLOR COLERIDGE
The Rime of the Ancient Mariner

In late May of 1768, Lieutenant James Cook, a young officer in the Royal Navy of King George III of England, received an unusual assignment from the British Admiralty. He was to sail to the South Pacific on HMS *Endeavour* to make astronomical observations of the planet Venus as it passed directly between the Sun and Earth, an orbital event that would take place in early June of the following year. Such a passage, known as a transit of Venus, eclipses a very small circular area on the face of the Sun that appears like a shadow moving across the solar disk. This astronomical phenomenon offered a method of estimating the distance between the Sun and Earth, by simultaneous observations of the moving dark spot from different points on Earth. Cook was to make his observations on the island of Tahiti in the Pacific Ocean, on the opposite side of the globe from England. The ostensible motivation for this undertaking lay in the suggestion that an accurate determination of the Earth-Sun distance was important for reliable navigation at sea.

The complexities of the motions of Earth and Venus about the Sun make transits relatively rare events, coming in pairs separated by eight years, but with more than a century separating one pair from the next. After the 1761/1769 pair, the next chances to observe a transit would come in 1874/1882 and 2004/2012. Cook had been selected for this scientific undertaking because of his skills in surveying and charting, honed a decade earlier on the St. Lawrence River, during the Seven Years' War between Britain and France for control of the territory that would become Canada.

Endeavour was a small ship, just a little longer than a modern railway coach, but home to eighty-five seamen and another dozen officers and accompanying naturalists, plus their equipment, water, provisions, and grog. The voyage from England to Tahiti followed a route south through the Atlantic, around Cape Horn at the tip of South America, and thence west into the Pacific to Tahiti. The full journey totaled roughly twelve thousand miles, equivalent to about half the distance around the globe. Under sail it took almost exactly eight months to reach Tahiti, including provisioning stops in Madeira and Rio de Janeiro, and some specimen collecting in Tierra del Fuego.

Cook was meticulous about the health of his crew, as the scourge of scurvy was already well known on long voyages. He knew that diet was important to health, and he carried an ample supply of sauerkraut to ward off scurvy. The crew, had they known of it, would have lobbied hard for the anti-scorbutant that Dutch sailors preferred: white wine. It is not clear whether Cook was aware of the prophylactic powers of wine, but he clearly knew the perils of having alcohol-incapacitated seamen. Christmas Day of 1768, celebrated off the coast of Patagonia, was marked not by religious services, but by a crew pursuing total inebriation. One of the naturalists remarked that they were lucky the Christmas winds were light.

Endeavour arrived in Tahiti in mid-April of 1769, in ample time to prepare for the astronomical observations. Cook selected a place to conduct the measurements—on a sandy beach not far from the present-day city of Papeete. He called the place Point Venus. When I visited Papeete a few years ago I was keen to see this famous scientific spot, but I worried that in the more than two centuries since Cook was there, the place might have lapsed into nothingness. I asked a taxi driver if he had ever heard of Point Venus. Yes, he replied, he knew it well. Skeptical that it would be so easy to find this historic place, I queried him further. Yes, yes, he knew the spot. So I asked him to take me there, and fifteen minutes later we arrived. It was Point Venus all right—but today well known as a popular nudist beach! Incidentally, there is also a small monument to Captain Cook's 1769 visit.

WHILE THE TRANSIT of Venus was the announced scientific rationale for this voyage, Cook's sailing orders from the Admiralty had another component, designated as secret and not to be opened by Cook until he was at sea. These orders addressed *Endeavour*'s assignment after the astronomical observations had been completed. They revealed that Cook was to search for Terra Australis Incognita, a hypothetical southern continent that had supposedly been dimly sighted in high southern latitudes by earlier mariners.

The notion of a southern continent had been promoted through philosophical and aesthetic arguments by Aristotle and later Ptolemy two millennia before the Age of Exploration. They believed that symmetry and balance were inherent characteristics of the natural world, and that Earth, as a natural object, must surely display these qualities. Such beliefs required the existence of landmasses in the Southern Hemisphere to balance the extensive landmasses of the Northern Hemisphere.

Not long after the transit was over—only six hours after it began—Cook took *Endeavour* southward in search of a southern continent.

Sailing south in the peak of the Southern Hemisphere winter quickly led to cold encounters with widespread sea ice, and it did not take long for Cook to realize that it was not the right season for a course into high latitudes. In September he headed west and encountered today's New Zealand. He proceeded to circumnavigate and chart the coastlines of both the North and South Islands, demonstrating that they were not a large southern continent, as had been surmised by earlier explorers. The return to England was by way of Australia, where *Endeavour* narrowly avoided disaster on the Great Barrier Reef, then onward to the East Indies, where several crew contracted malaria, and around Africa to the Atlantic, before heading north on the last long leg home. In the Atlantic he encountered some American whalers, and stopped to get news of the last three years—he learned that Europe was, for a change, at peace. Cook arrived in England in the summer of 1771, with no sighting of Terra Australis Incognita to report.

The return of *Endeavour* was celebrated and acclaimed widely, but the focus was not on Cook, the modest master of the vessel. In the limelight was the young patrician naturalist Joseph Banks, well versed in manipulating the press to his advantage. Within just a few weeks, Banks had worked up a frenzy of public adulation in the press that culminated in his announcement that there would soon be a second voyage of exploration and scientific discovery, under his leadership. Incidentally, Banks would insist that Cook undertake the maritime duties, and there was little Cook could do to decline. Within a month of his returning home after an absence of three years, Cook was already planning the next sailing. His wife, Elizabeth, was not too pleased.

In 1772, by then promoted to captain, the rank by which he is best remembered, Cook sailed again for the Southern Ocean aboard a new ship, HMS *Resolution*, once again in search of Terra Australis Incognita. On this voyage he headed toward the Pacific by turning east around Africa into the Indian Ocean, and pushing to ever higher southern latitudes as ice conditions would permit. In 1773 he crossed the Antarctic

Circle[1] three times, at longitudes 40° east, 140° west, and 105° west; each time he encountered impenetrable ice, and came away without sighting a southern continent.

His eastward course across the South Pacific, never far from the ice, brought him to the southern tip of South America just as 1774 ended. Early in the new year, he sailed eastward into the South Atlantic, and discovered South Georgia Island, a banana-shaped glacier-striped island that, at first sighting, he thought might be the long-sought southern continent. But when the distal tip of the banana came into view, he knew it was just an island. He named it Isle of Georgia, in honor of King George III. Continuing eastward, Cook reached the cape of southern Africa, intersecting his path around Africa three years earlier. He had now circumnavigated the globe in the southern high latitudes, seldom very far from the edge of the ice. Cook noted in his journal[2]:

> I had now made the circuit of the Southern Ocean in a high latitude and traversed in such manner as to leave not the least room for the possibility of there being a continent, unless near the pole and out of reach of navigation. . . . The greatest part of this Southern Continent (supposing there is one) must lie within the Polar Circle where the sea is so pestered with ice that the land is thereby inaccessible. . . . I can be bold to say that no man will ever venture farther than I have done, and that the lands which may lie to the south will never be explored. Thick fogs, snowstorms, intense cold and every other thing that can render navigation dangerous one has to encounter, and these difficulties are

1. The polar circles in the Arctic and Antarctic are at 66.6° north and south latitudes, respectively. The polar circles define the areas around the poles that experience round-the-clock daylight at least one day each summer, and total darkness at least one day each winter. The number of days of such illumination (or the lack thereof) increases toward the poles. The distance from the polar circles to the poles is 23.4°, equal to the tilt of Earth's rotational axis away from being perpendicular to the plane of Earth's orbit about the Sun.
2. This extract from Cook's journal is from *Captain James Cook*, by Richard Hough (New York: W. W. Norton, 1994).

> greatly heightened by the inexpressible horrid aspect of . . . a country doomed by nature never once to feel the warmth of the sun's rays, but to lie for ever buried under everlasting snow and ice.

Cook had clearly disproved the hemispheric "balance" of landmasses postulated by Aristotle, but he demonstrated symmetry of a different type, symmetry not of land but of ice. He had shown that there was a daunting ice barrier in the high latitudes of the Southern Hemisphere, similar to that encountered in the Arctic. His predictions about the inaccessibility of the polar latitudes in the South, however, did not stand. In the early nineteenth century several sailing ships did indeed sight the Antarctic continent.

In 1838, just a little more than a half century after the founding of the nation, the United States sent an expedition to the South Pacific and Antarctic, formally called the United States Exploring Expedition of 1838–43, but colloquially known as the "U.S. Ex Ex." The expedition was commanded by Lieutenant Charles Wilkes, a naval officer, but was well staffed with scientists, the best known of which was the noted biologist and geologist James Dwight Dana. In early 1840 the expedition reached the icy barrier along the coast of Antarctica just at the Antarctic Circle, two thousand miles south of Australia. Wilkes traced the coastline for more than fifteen hundred miles, equivalent to the distance from Boston to Miami. Proof that this extensive terrain was indeed a continent would come later, but clearly the U.S. Ex Ex had encountered a big and continuous landmass.

THE SEVENTH CONTINENT

The symmetry of ice in both the northern and southern high latitudes sometimes conveys a false impression that Earth's polar regions are really quite similar. The presence of ice, however, actually masks more fundamental differences between the north and south polar regions. The Arctic and Antarctic have been described as being "poles apart," of course geographically, but also in many other characteristics. The South Pole lies well within the continent of Antarctica, some 850 miles inland from, and 10,000 feet above, the nearest coastline. The North Pole, by contrast, is located in the Arctic Ocean, with the seafloor 14,000 feet below and the closest coast some 450 miles away. Both poles are set in ice, but the thickness of the ice is very different. Beneath the South Pole lies more than 10,000 feet of ice, whereas the North Pole sits on a thin 10- to 20-foot sheet of frozen ocean water, give or take a few feet. The ice in both settings is on the move, but at very different speeds—at the South Pole the ice slips slowly over the pole at a glacial pace of about 30 to 40 feet per year, whereas the sea ice of the Arctic is swept along by wind and currents at an average speed of about 3 to 4 miles per day.

Size-wise, Antarctica is a typical continent—smaller than Asia, Africa, North America, and South America, but larger than Europe and Australia. And it shares many geological characteristics with the other continents. The large-scale architecture of all continents is similar to that of icebergs—continents are composed of rocks, such as granite, that are less dense than the rocks that make up the floors of the surrounding ocean basins. Just as ice floats in water, with some ice above but most below the water's surface, continental rocks "float" in rocks of greater density, and stand a bit higher than the rocks in which they are immersed. The average elevation of the continental surface is some three miles above the ocean floor, but the low-density rocks of the continents extend more than twenty miles into the Earth, a continental "root" not unlike the submerged portion of an iceberg in the ocean.

As in the other continents, the Antarctic rocks show the telltale characteristics of a long and complex geologic history—a wide range of ages, from ancient Precambrian crystalline rocks to very young unconsolidated glacial deposits. The rock types include the common rock categories—igneous, sedimentary, and metamorphic—and in typical proportions. The Antarctic continent has mountain ranges such as the Antarctic Peninsula, which is really just an extension of the Andes of South America, and the Transantarctic Mountains, which snake across the continent from the Weddell Sea to the Ross Sea. Antarctica almost certainly has its share of mineral deposits, although none is exploitable, at least for now, because of the extreme environment. Antarctica is, however, unique in one important characteristic—its location astride the South Pole. Virtually all of Antarctica lies within the Antarctic Circle, and more than three quarters of its area lies at latitudes greater than 70° south.

How and when did Antarctica come to the South Pole? One might be tempted to ask, "Hasn't it always been there?" but there is ample geologic evidence to indicate that it has not. Sedimentary rocks of Mesozoic age along the Antarctic Peninsula show beautiful fossilized tropical ferns, and Paleozoic-age coal seams in the Transantarctic Mountains reveal well-preserved low-latitude vegetation. No, Antarctica was not always at the South Pole—it came there from somewhere else, and fairly recently, geologically speaking.

At the beginning of the Jurassic period, some two hundred million years ago, the terrain that was to become Antarctica was part of a super-continental assemblage called Gondwanaland, an enormous landmass that also comprised the eventual continents of South America, Africa, and Australia, as well as smaller fragments including Madagascar, New Zealand, and India. Gondwanaland itself had been assembled only one hundred million years earlier, during the closing stages of the Paleozoic era. Following its assembly from predecessor continental terrains from around the globe, this composite landmass received deposits of a unique and remarkably widespread sequence of rock formations, and saw the evolution of a cosmopolitan fauna and flora. Geologists and paleontologists eventually recognized this rock sequence with its contained fossils as the Gondwanaland signature—the key to recognizing the full extent of Gondwanaland.

About 170 million years ago, the forces of plate tectonics began to dismember Gondwanaland and disperse the pieces. Just as sea ice glides slowly over the surface of the high-latitude oceans, so also do large segments of Earth's rocky outer shell drift slowly over the globe, mobilized by forces from within the planetary interior.

The continental dispersal created a new geography in the Southern Hemisphere. Within Gondwanaland, Antarctica was originally situated at about 40° south, and governed by a temperate climate very similar to that characteristic of the continental United States today—neither polar nor tropical. Widespread forests and marshes of the time were eventually compressed into the coal beds found today in the Transantarctic Mountains.

The separation of Antarctica, Madagascar, India, and Australia from Africa, and from one another, created a gap that became the modern Indian Ocean. A little later, the departure of South America from Africa created the South Atlantic Ocean. India went its separate way northward across the equator, eventually to collide with southern Asia to create the Himalaya mountain range. Australia and Antarctica were carried southward.

The defining tectonic events for Antarctica, the events that make it unique, came around thirty to forty million years ago. Australia parted company with Antarctica and headed north, leaving Antarctica to enjoy the pole alone. And as Antarctica slipped farther south, the Andean link between South America and the Antarctic Peninsula was stretched and then broken, opening a six-hundred-mile-wide oceanic chute known today as the Drake Passage. Antarctica was then totally surrounded by the Southern Ocean, a ring of water around the globe at 60° south. The prevailing wind at that latitude blows from west to east, and it sets up an ocean current, the Antarctic Circumpolar Current, that circles Antarctica relentlessly.

THE ISOLATION OF ANTARCTICA

The climatological impact of the west-to-east circumpolar current has been profound. With virtually no flow in a north–south direction, the current inhibits mixing of the cold Southern Ocean with warmer waters of the Atlantic, Pacific, and Indian oceans. Unlike the Arctic region, which receives tropical warmth via the northward-flowing Gulf Stream of the Atlantic Ocean, the Antarctic is climatologically isolated by this circulatory girdle. In the Arctic, the port of Murmansk, in Russia, remains ice-free throughout the year, even though it is located well north of the Arctic Circle. By contrast, in the Antarctic there is not a single place south of the Antarctic Circle that is free of winter sea ice.

There are many definitions for the boundary of Antarctica. The continental coast defines the geographic boundary, the margin of the Antarctic tectonic plate delimits the geological boundary, and the 60° parallel of south latitude marks the political boundary governed by the Antarctic Treaty. But the climatological boundary, the boundary that makes Antarctica unique, is defined by the abrupt north-to-south transition from warmer temperate-zone water to frigid polar water within the Antarctic Circumpolar Current. It is not unlike the "marriage of the waters" in Brazil, at the confluence of the Rio Negro and the Amazon. There the dark water of the Rio Negro flows side by side with the tan, muddy waters of the Amazon, but after a few miles of getting acquainted, they mix together and become one. In the Antarctic, however, the winds and currents maintain the large temperature differences, and prevent a mixing of the waters. They flow side by side in a courtship never consummated—a marriage (surely not the first) thwarted by frigidity. This climatologic boundary is known as the Antarctic Convergence.

The crossing of the Convergence is marked by a drop in the temperature of the seawater of nearly ten Fahrenheit degrees, and the air temperature chills accordingly. Fog is an occasional visible marker, and the appearance of icebergs, first a few and later many, raises the navigational ante as ships penetrate farther south. The radar on a ship's bridge slowly becomes speckled with reflections from the bits and pieces of ice. Soon thereafter, large floating "islands" of ice appear. The continent is not yet visible, but it is very clear that you have arrived in the Antarctic.

When you finally reach the continent, your feelings are overtaken by the pristineness and simplicity of the landscape. Mountains rise from the sea, draped entirely in white. Large serpentine glaciers a mile across wind through the landscape, apparently static, but in reality slithering slowly downward from the heights—giant conveyor belts delivering huge blocks of ice to the sea. The seas surrounding the continent are clogged with titanic icebergs, of extraordinary size and architecture. The vista is powerful, yet quietly serene. Aboard *Nimrod* in early 1908, Ernest Shackleton described his arrival:

> As far as the eye could see . . . the great white wall-sided bergs stretched east, west, and south, making a striking contrast with lanes of blue-black water between them. A stillness, weird and uncanny, seemed to have fallen upon everything when we entered the silent water streets of this vast unpeopled white city.[3]

The landscape is vast but also deceptive—it is without most of the visual cues that attach scale, distance, and dimension to the natural world elsewhere. Indeed, the simplicity emerges from what the landscape is free of. There are no people; no buildings or construction cranes; no telephone poles or microwave towers; no roads, cars, trucks, or snowplows; no cultivated fields or irrigation circles; no airplanes overhead; no billboards, junkyards, or trash mounds. And the natural world is also limited—no bushes, hedges, trees, or forests; no tulips, sunflowers, lupines, or forsythia; and no wolves, deer, moose, or caribou.

The aural "landscape" is also very different. There are no industrial sounds; no deep rumble of diesel engines; no hissing, humming, whining, or thumping; no blaring music; no honking horns or sirens. The ubiquitous sounds of the Antarctic are those of wind, water, and ice. Winds whistle at fifty, sixty miles an hour, and waves crash with great thuds on beaches of volcanic rock, or against rocky or icy cliffs. Glaciers creak and crack as they inch their way through rocky valleys. And superposed on the inanimate

3. Ernest Shackleton, *The Heart of the Antarctic* (New York: Signet, 2000), 44.

sounds are those of the wildlife—whales spouting, seals belching, penguins calling. Petrels, gulls, and albatross ride the wind in almost total silence. This is truly "the world without us,"[4] a frozen part of the Garden of Eden that has been off limits to us for most of human history.

The colors of the Antarctic are unlike colors elsewhere. Whereas green is the signature color of well-watered vegetation everywhere, and reds, yellows, and tans paint Earth's deserts, Antarctica specializes in black, white, and blue. The rock is mostly black and the snow white. Glacial ice is white at the surface, but deep brilliant blue where crevasses and fissures reveal the interior. On a cloudy day, the deep sea is dark, and when the Sun shines brightly, the ocean appears a very deep blue. In brilliant sunshine the sky is a perfect sky blue, and when clouded over, it is a blank sheet of low-hanging gray. In deep fog a three-dimensional gray shroud settles in, completely disrupting one's sense of orientation and distance.

The Sun in the Antarctic summer is never far above or far below the horizon—it simply rides around the horizon, offering an ever-changing azimuth of illumination that casts pink hues and slowly changing long shadows that sweep across the landscape. The polar circle cuts through the Antarctic Peninsula about halfway through its lineal extent. South of the circle are long stretches of summer, when the Sun never sets, and north of that line the Sun dips just below the horizon for an hour or two, creating a very long "sunset" of delicate pinks, before returning to view and offering direct illumination once again.

Wind is erratic. A transition from total calm to gale-force winds can occur unexpectedly, the result of very cold and dense air suddenly spilling off highlands and roaring through valleys. These winds, called katabatic winds, are the atmospheric equivalent of a flash flood. They come without announcement, bluster through with abandon, and are gone within minutes. They can drive inattentive ships into rocks and flatten humans caught unaware.

---

4. An image of the world before humans put their signature on it, and as it would return to a natural state after a hypothetical departure of humans from it, can be found in Alan Weisman's *The World Without Us* (New York: St. Martin's Press, 2007).

But nothing quite matches the special experience of getting up close and personal with big icebergs. Conveying the scale of bergs requires reference to something you can envision, so let's start with a ship of the type that has brought me to the Antarctic several times—an oceangoing vessel more than four hundred feet long and almost one hundred feet high. When such a ship positions itself in the lee of a middling iceberg, the vessel is dwarfed, silhouetted against a floating ice island that easily exceeds the ship in both length and height. The ship becomes a miniature, not in a bottle, but in a vast field of icebergs. A ship that would fill a football stadium does not quite measure up.

Icebergs generally come either from a glacier discharging great chunks of ice into the sea, or from the margins of a floating ice shelf. The distinction is artificial, however, because the ice shelves themselves are fed by glaciers. But the shelves tend to lose the irregularity of the glacial ice that feeds them, eventually to exhibit a flat upper surface like a tabletop. When a shelf launches an iceberg through breakup or break-off, the berg retains the flat top (at least for a while), and accordingly is identified as a tabular berg. The chunks that calve from the snout of a valley glacier are much more irregular, depending on the extent of crevassing that develops in the glacier as it creeps through its valley toward the sea.

Once an iceberg is in the sea, wind and water take over its destiny. Afloat, a berg will bob up and down like a giant cork, rising, falling, swaying, and tilting in slow motion. Sometimes a floating berg will break in two, and for a few minutes each offspring berg will slowly rock and roll in the sea, seeking a new equilibrium that places its center of gravity in a stable position below the surface. Sometimes this process leads to a complete overturning that brings the formerly submerged portion of the berg to the surface. If a berg is blown into shallower water, it may run aground and await a high tide for relaunching. Or it may sit there for years, slowly being diminished by the pounding of waves. Wave erosion creates a "waterline," where the ice and the sea surface meet; some bergs display many waterlines at different elevations and intersecting angles, telling a history of grounding and refloating, and of re-equilibration following a breakup.

The sculpting of icebergs by the elements has always fascinated observers, and opened their imaginations to interpreting the myriad shapes. Icebergs are to the polar imagination what cloud forms are to people elsewhere. Frank Worsley, the captain of Sir Ernest Shackleton's ship *Endurance*, offered this description of a field of Antarctic icebergs:

> Great fragments and hummocks of very old floes, worn, broken down, and melted into all sorts of grotesque and wondrous shapes, were heaving, bowing, curtseying, and jostling on the long westerly swell. . . . Castles, towers, and churches swayed unsteadily around us. Small pieces gathered and rattled against the boat. Swans of weird shape pecked at our planks, a gondola steered by a giraffe ran foul of us, which amused a duck sitting on a crocodile's head. Just then a bear, leaning over the top of a mosque, nearly clawed our sail. An elephant, about to spring from a Swiss chalet on to a battleship's deck, took no notice at all; but a hyena, pulling a lion's teeth, laughed so much that he fell into the sea, whereupon a sea boot and three real penguins sailed lazily through a lovely archway to see what was to do, by the shores of a floe littered with the ruins of a beautiful white city and surrounded by huge mushrooms with thick stalks. All the strange, fantastic shapes rose and fell in stately cadence, with a rustling, whispering sound and hollow echoes to the thudding seas, clear green at the water line, shading to a deep dark blue far below, all snowy purity and cool blue shadows above.[5]

WHAT LURED PEOPLE into the polar ice? Fame, glory, adventure, and career advancement were important motivations for explorers and naval

5. F. A. Worsley, *Shackleton's Boat Journey* (New York: W. W. Norton, 1977), 104–5.

officers, but fortune, territory, and geopolitical power were what the commercial and national sponsors of exploring expeditions generally hoped for. By early in the twentieth century all the land surrounding the Arctic Ocean was politically attached to either Russia, the United States, Canada, Denmark, or Norway, and the ocean itself, mostly covered with year-round sea ice, was at that time not a sufficiently attractive commercial target to promote international tensions. However, the situation in the Antarctic was different.

SLICING THE ANTARCTIC PIE

Although at the end of the nineteenth century neither the North nor South Pole had been reached, the route to the South Pole was over land, and in that heyday of imperialism, "vacant" land invited territorial claims. The Berlin Conference of 1884 had partitioned Africa for the benefit of the European powers; France, Germany, Belgium, Portugal, Great Britain, Italy, and Spain imposed colonial governments on more than 95 percent of the African territory.

Antarctica was unclaimed land. Although it was not an inviting place to establish colonies of settlers, nor seen as a great opportunity to enrich national treasuries and privileged royalty, it nevertheless offered the prestige factor of adding more pink or lavender or green to imperial world maps. And it had some strategic military value in terms of control of the Drake Passage connecting the Atlantic and Pacific, a value that was diminished after the opening of the Panama Canal in 1914.

By the end of the first decade of the twentieth century, most of the European nations that had set up colonial regimes in Africa were active in exploring and exploiting the coast of Antarctica, but they were joined by Norway, Sweden, and the Southern Hemisphere nations of Australia, New Zealand, Chile, and Argentina. Both Norway and Great Britain had penetrated the interior of Antarctica as well, reaching the South Pole in

December 1911 and January 1912, respectively. Britain initiated the claiming of Antarctic territory in 1908, even before reaching the pole. World War I intervened briefly while the European powers fought with one another for imperial supremacy, but over the next twenty-five years, Australia, New Zealand, France, Norway, Chile, and Argentina announced Antarctic territorial claims. These claims were typically drawn as "pie slices," with the center of the pie at the South Pole. The claims of Chile, Argentina, and Great Britain, however, inconveniently overlapped with one another, and as World War II came to a close in the Northern Hemisphere—the seeds of conflict had been planted in the territorial claims in Antarctica.

The end of World War II also saw the emergence of a new global power structure, the preeminence of the United States and the Soviet Union, and the nascent cold war between them. The United States had been active in Antarctica—from the U.S. Ex Ex presence in 1840 to the geological explorations and 1929 flight over the South Pole by Commander Richard Byrd from his Little America base on the Ross Ice Shelf. After World War II the United States returned to Little America to conduct Operation High Jump, a military exercise of 4,700 troops, 12 ships, and 9 planes.

The Soviet Union, however, was a newcomer to the Southern Hemisphere. Imperial Russia had sponsored Fabian Gottlieb von Bellingshausen's 1819–21 circumnavigation of the globe, which included a sighting of Antarctica in 1820, but nothing thereafter. The decade following World War II saw the cold war take full form—the Berlin Airlift, the Korean War, and the nuclear weapons race. The Soviets were asserting themselves everywhere, and soon, perhaps not surprisingly, the cold war came to the cold continent. The Soviet Union rejected the notion of national territories in Antarctica, and in 1950 made its position very clear when it stated that it would not recognize as lawful any decisions taken on Antarctica without its participation. The growl of the Red Bear echoed across the white continent.

The United States also rejected all existing land claims, and to emphasize the point it set up a research station at the South Pole. By "occupying" the South Pole, at the center of the continental pie, the United States could then symbolically claim control in all directions, over the full 360° of azimuth radiating outward from the pole. But it was only symbolism to make a point; the nominally non-imperial policy of the United States had long been to eschew claims of territory in the Antarctic.

In the face of the contentious overlapping claims of Argentina, Chile, and Britain on the Antarctic Peninsula, Chile in 1948 proposed a five-year suspension of sovereignty issues, and urged instead tripartite scientific collaboration. In the following year, the three nations signed a treaty barring military vessels south of latitude 60°. But by 1952, Argentina had built a base at Hope Bay on the peninsula, only a few hundred yards away from a British base that had partially burned a few years earlier. When later that year the British returned to rebuild their base, the Argentines fired warning shots over the heads of the British reconstruction crew. These were the first shots fired in hostility in Antarctic history, and did not augur well for a peaceful future in Antarctica. Britain brought in the Royal Marines to protect the reconstruction.

The deteriorating political situation in Antarctica invited a more sober alternative, one that would defuse the incendiary incident at Hope Bay and perhaps prevent what was apparently looming near—an inevitable conflict of national interests throughout the continent. Interested nations discussed ways to make Antarctica a continent for science, and a continent for peace. Thus was born the concept of what would become known as the International Geophysical Year of 1957–58.

INTERNATIONAL POLAR YEARS

The idea for an international scientific year focusing on the high latitudes was not altogether new. The first International Polar Year (IPY) occurred in 1882–83, just before the imperial knife was readied for carving up

Africa.[6] This multinational cooperative research venture at latitudes beyond the polar circles was a recognition that much of atmospheric circulation and accompanying meteorology were affected strongly by the polar regions, and that navigation by magnetic compass would benefit greatly from investigations near the magnetic poles. Moreover, as was well known to all, working in the polar regions was difficult, dangerous, and costly, and therefore nations were willing to undertake cooperative ventures to share both risks and costs, and to keep a geopolitical eye on one another. Most of the research expeditions of this first IPY were to the Arctic, but three went to Antarctica. The second IPY took place a half century later, during the Great Depression, again focusing principally on the Arctic. A third IPY had deployments occurring throughout 2007–9.

THE INTERNATIONAL GEOPHYSICAL YEAR AND THE ANTARCTIC TREATY

The International Geophysical Year (IGY) of 1957–58 was an extraordinary scientific and geopolitical success. Perhaps it was because of the urgency at that time to find a way to avoid repeating the many geopolitical mistakes of the past. Or perhaps it was simply that there was a great deal of scientific interest in the polar regions, and new logistical capabilities and new scientific technologies made 1957–58 a perfect window of opportunity. Nothing symbolized the new technology more than the launching of the first artificial satellites to orbit Earth—the Soviet *Sputnik 1*, in October 1957, and the United States' *Explorer 1*, four months later. And nothing characterized the spirit of scientific cooperation better than the establishment of an international data center, where observations from all the national expeditions were to be archived and shared.

6. The close juxtaposition in time of the International Polar Year and the Berlin Treaty displayed a tenuous thread of consistency—cooperation among the nations of Europe when it served their national self-interests.

Most nations that participated in the IGY were delighted with its outcome, and wanted to perpetuate the science and cooperation model of activity in Antarctica. The principles of the IGY were translated into a diplomatic document known as the Antarctic Treaty, first adopted in 1959 and ratified in 1961 by the United States, the USSR, the United Kingdom, and nine other nations with active research programs on the white continent.

The treaty addressed many issues, but a few stand out clearly. The first article declared Antarctica a continent for peace, and laid out provisions to ensure that the continent would remain a demilitarized region. The second article declared Antarctica a continent for science, free and open everywhere for scientific investigation and cooperation. The treaty defused the conflicting territorial claims simply by saying that maps could be drawn however nations might wish, but no enforcement of claims or restrictions on travel would be allowed. Important wildlife conservation protocols were later adopted, as was a moratorium on exploration and exploitation of mineral resources that extends to the year 2043.

The treaty, reaffirmed in 1991 and today with more than forty signatories, has shown how shared governance by mutual consent has shaped a new style of international relations. That Antarctica stands alone as a continent for peace, multinational cooperation, scientific research, and non-exploitation is a remarkable outcome of the IGY and the subsequent Antarctic Treaty.

"GOVERNANCE" IN THE ARCTIC

As I note earlier, land claims in the Arctic never became quite the issue that they did in the Antarctic. The countries surrounding the Arctic Ocean had more or less well-defined boundaries, and "ownership" of the few islands situated beyond obvious national affiliations was adjudicated through treaties. The question of how far national sovereignty extended into the adjacent Arctic Ocean was essentially moot because of the great difficulties the perennial sea ice imposed on resource exploitation. The relevant international law on this subject is embodied in the United Nations Convention on the Law of the Sea, to which the United States is not a signatory.

But in the mid-twentieth century, if the Arctic Ocean had no immediate commercial significance, it very much had military importance, and both the Soviet Union and the United States recognized this. While the International Geophysical Year had offered the promise of peaceful coexistence, at least in Antarctica, the cold war continued elsewhere. In 1958, the U.S. nuclear submarine *Nautilus* set out from Seattle on a nominally routine cruise in the North Pacific, but as with Captain Cook and HMS *Endeavour* in 1768, *Nautilus* also had secret orders: disappear beneath the surface of the North Pacific, and then enter the Arctic Ocean clandestinely through the Bering Strait. *Nautilus* was to explore and chart the topography of the Arctic Ocean basin, and make observations of the sea ice thickness overhead. And in another display of late-1950s scientific and engineering prowess—artificial satellites were the first, just a few months earlier—*Nautilus* broke through the sea ice and surfaced at the North Pole, sending home the terse message "Nautilus Ninety North." In effect, the appearance of *Nautilus* at the pole was to announce to the world that no place in the oceanic domain was beyond the reach of American naval power. William R. Anderson, the skipper of *Nautilus*, brought a piece of Arctic ice home as a souvenir for Admiral Hyman Rickover, the curmudgeonly father of America's nuclear submarine fleet.

The Soviets also recognized the military significance of the Arctic Ocean. If nothing else, it was a well-camouflaged shortcut that could bring the contiguous United States quickly within range of submarine-launched missiles. Over several decades the submarines of the cold war powers played cat-and-mouse with each other, and carefully monitored submerged traffic beneath the sea ice cover of the Arctic Ocean. A by-product of this activity was an ever-increasing archive of scientific information about the Arctic: the topography of the ocean floor, the thickness of the sea ice from place to place, the nature of the magnetic field near the north magnetic pole, and the speed of sound transmission through the oceanic waters.

But it was not a single archive of scientific observations that was being compiled—there were two, one American and one Soviet. Detailed maps and charts of the Arctic bathymetry could reveal potential hiding places for submarines, and knowledge of the magnetic field could help military intelligence officers assess how the magnetic signature of a submarine could be suppressed or disguised. The United States and the Soviet Union were in effect conducting parallel and redundant geophysical surveys of the Arctic marine environment.

The cold war ended with the dissolution of the Soviet Union in late 1991. By the end of 1992, Boris Yeltsin and Viktor Chernomyrdin were occupying the offices of president and prime minister of the Russian Federation, respectively. In the United States, Bill Clinton was elected president and Al Gore as vice-president in 1992. The new leadership in both countries presented new opportunities for cooperation. At a summit meeting the next year, Clinton and Yeltsin established a bilateral commission, headed by Gore and Chernomyrdin, to promote cooperation between the former adversaries of the cold war. The initial focus of the commission was on space, energy, and high technology, but soon encompassed health, agriculture, science, and the environment as well. Within a year the two countries had signed an agreement that addressed environmental issues in the Arctic.

Gore and Chernomyrdin both recognized that each country possessed geophysical data about the Arctic Ocean that no longer offered military advantage, because each country had independently acquired the same data. In a remarkable turnabout from the cold war posture, they decided to release the data to the international science community. Depth soundings, water temperature and salinity measurements, ice thickness and ocean current maps, meteorological observations and much more would come out of security vaults and be placed in the public domain. The result was the publication of the U.S.–Russian Atlas of the Arctic Ocean in 1997. Vice-President Gore remarked that "some of science's most sought-after data about our environment has literally 'come in from the cold' . . . a great portal of knowledge has swung open."[7]

The Gore–Chernomyrdin vision was prophetic. The information released, acquired between 1948 and 1993, has provided the historical baseline with which we compare changes taking place today in the Arctic. It is because of this data that we can recognize the seriousness of the decline in Arctic summertime sea ice, a seasonal loss that has accelerated dramatically in the early years of the twenty-first century. And the spirit of international cooperation blossomed—the 2004 Arctic Coring Expedition (ACEX) comprised scientists and ships from a dozen nations, including my University of Michigan colleague Ted Moore, a marine geologist. ACEX returned with drill cores from the bottom of the Arctic Ocean that revealed fifty-five million years of fascinating high-latitude geological history[8] and changing climate. Fifty-five million years ago the global climate was very warm, a condition brought about by a release into the atmosphere of the greenhouse gas methane, long sequestered beneath the ocean floor. It was the last time the entire planet was free of ice.

Currently, however, international attitudes about the Arctic are once again turning colder. The fast-diminishing sea ice in the Arctic Ocean has opened the possibility of easy access to vast reaches of the ocean that have been inaccessible for millennia or longer. Nations surrounding the Arctic Ocean are now imagining the possibilities of petroleum and natural gas, trade routes and fisheries. There is renewed interest in novel interpretations of the Law of the Sea as a vehicle of governance in the Arctic. This newly developing geopolitical turbulence will only be amplified by the fast-approaching disappearance of summer sea ice in the Arctic over the next few decades.

---

7. "Arctic Breakthrough," *National Geographic* 191, no. 2 (1997): 36–57.

8. T. C. Moore and the Expedition 302 Scientists, "Sedimentation and Subsidence History of the Lomonosov Ridge," in J. Backman, K. Moran, D. B. McInroy, L. A. Mayer, and the Expedition 302 Scientists, *Proc. IODP, 302* (Edinburgh: Integrated Ocean Drilling Program Management International, 2006), doi: 10.2204/iodp.proc.302.105.2006.

TOURISTS COME TO THE POLAR ICE

As I describe in chapter 6, we humans have left our mark on the land, air, and water everywhere we have settled. As our numbers and energy usage have grown dramatically, the human footprints on the globe are nearly ubiquitous. But if ever there were places seemingly unaltered by people, one would think first of the icy polar regions—Antarctica in the South, and Greenland and the Arctic Ocean in the North. Throughout the eighteenth, nineteenth, and early twentieth centuries, the high latitudes were accessible only to explorers, whalers, sealers, scientists, and naval flotillas, with many expeditions a blend of these differently motivated purposes. What they all had in common were the facts that the polar regions were hard to reach, inhospitable in the extreme, dark half the year, and dangerous.

But such hazards did not discourage people with a sense of adventure (and a willingness to pay) from joining expeditions. Apsley Cherry-Garrard's application to join Robert Falcon Scott's *Terra Nova* expedition to Antarctica in 1910 was at first rejected, but when Cherry-Garrard contributed £1,000 (about $100,000 today) to the expedition, he was allowed to come along.

Access to the polar regions began to change in the 1960s, with the advent of transportation that enabled tourists and adventurers to reach high latitudes without benefit of military transport, scientific logistical support, or resource-driven commercial enterprises. The first ship custom-built for expeditionary tourism was the MS *Lindblad Explorer*, the vision of Lars-Eric Lindblad, a Swedish American who saw the business potential of tourism in the remote places of the world. Launched in 1969, the *Lindblad Explorer* took adventurous tourists to both the Peninsula and the Ross Sea sectors of Antarctica, through the Northwest Passage of the Canadian Arctic from the Atlantic Ocean to the Bering Sea, and to Svalbard, the Norwegian island at 78° north, where the Atlantic Ocean meets the Arctic Ocean.

The *Lindblad Explorer* was painted bright red, and became known as the "Little Red Ship." *Explorer* was not an icebreaker, but she had an ice-rated double hull that enabled her to move slowly through loose sea ice, gently nudging the ice fragments aside. At capacity *Explorer* could carry around a hundred passengers, and over the Antarctic summer season she could provide the Antarctic experience to around a thousand visitors.

When I first went to Antarctica in 1990, it dawned on me that more people would watch a single football game in the University of Michigan Stadium—the largest stadium in America, with a capacity of about 110,000—than had ever been to Antarctica in all of human history. A decade later I could not say that anymore. Ships galore had begun to bring tourists to Antarctica—small ships, big ships, icebreakers—all recognizing the tremendous interest in seeing the splendors of the Antarctic before Earth's warming climate changed Antarctica forever. Today some fifty ships bring around forty-five thousand tourists to the Antarctic each year.

The most traveled touristic sea route to the Antarctic is from the southern tip of South America to the Antarctic Peninsula. This route is favored because it is the shortest route by far—only six hundred miles or so; the route from New Zealand is more than five times longer. This constriction in the Southern Ocean is called the Drake Passage, after Sir Francis Drake, a sixteenth-century privateer in the British Navy, well known for harassing Spanish vessels along the Pacific coasts of both North and South America.

Antarctica is only two sailing days from South America, but to reach it you must first cross the Drake Passage. Because of its narrowness and storminess, the passage has a well-deserved reputation for making a journey to Antarctica on occasion very uncomfortable, even in large modern ships with stabilizers. Forty-eight hours of rough seas is the price you must be prepared to pay to reach Antarctica—ten-foot swells, waves breaking over the bow and sending spray all the way to the navigational bridge. Cabin furniture can be sent careening, and crockery can slide off the dining room tables. There is an incessant thud as the ship, after being uplifted by a swell, comes crashing down on the sea; a thump, thump, thump as the turning propellers, temporarily lifted out of the water on the back of a big wave, carve their way back into the sea to resume their duty of pushing the ship southward. But there is the occasional surprise—sometimes the passage is so calm that the waters are affectionately called the Drake Lake.

Two days after departing South America, tourists reach the white continent. Blessedly the waters around the Antarctic Peninsula are sheltered and calm. Once in Antarctic waters, visitors can go ashore in small inflatable landing craft called Zodiacs, ten-passenger rubber boats powered by outboard motors—the vehicle of choice for both scientists and tourists in getting from place to nearby place in Antarctica. The landings are marine-style: leaping into shallow surf at the edge of the beach and scrambling ashore. They are appropriately called "wet landings," although knee-high rubber boots usually keep the visitors dry. Once ashore, the tourists visit penguin and seal breeding areas, hike up steep terrain to view the extraordinary landscape of ice caps, glaciers, and mountains. In the Zodiacs they tour close to calving glaciers and into iceberg "graveyards," sheltered bays where the wind drives many big bergs into temporary immobility. The Zodiacs offer unparalleled opportunities to become intimate with ice.

Many tourists are veteran world travelers who want to set foot on their seventh continent. For safety reasons, the rules of Antarctic tourism allow no more than one hundred people ashore at a given time. The task of ships avoiding one another at favorite destinations has grown into a scheduling and navigational challenge. Everyone who comes to the Antarctic imagines that they alone are having this once-in-a-lifetime experience. The last thing they want to see is another ship sitting at anchor in Paradise Bay, its passengers ashore enjoying a hike up to a special viewing point, or in Zodiacs exploring the face of a massive calving glacier. No, everyone wants a pristine Antarctica, unsullied even by the presence of others. Well before the tourist season begins, expedition leaders and ship captains submit requests to a clearinghouse for landing sites and times, much like booking admission times to popular museum exhibits weeks in advance. But in Antarctic waters, ever-changing wind, fog, and ice conditions frequently force last-minute shuffles in schedules. Advance planning is obligatory, but day-to-day improvisation is usually the reality.

Who guides tourists in the Antarctic? Aboard most ships there is a very small expedition staff of naturalists—ornithologists, marine biologists, geologists, glaciologists, historians, meteorologists, oceanographers—adventurous people who have gained Antarctic (or Arctic) experience, principally through scientific work. As the number of ships has grown, so has the need for naturalists familiar with the Antarctic. Today this small band of men and women probably number fewer than five hundred, distributed over some fifty ships for all or part of the season. Many have also spent years driving Zodiacs. The Antarctic setting can be a challenging one, with high winds, big waves, and bigger icebergs. Experience is at a premium—these folks are the ones who bear the responsibility of transporting tourists to the beach from ships at anchor offshore, and disembarking them safely at "unimproved" landing sites.

I have had the privilege of working with some of these remarkable people over the years. Russ Manning, affectionately known to colleagues as "Russ of the Antarctic," is a distant relative of Nanook of the North, with a wild mop of multicolor hair that is never covered by a hat no matter how bad the weather. Russ is a fifteen-year veteran of the Royal Marines who later commanded the British Antarctic Survey scientific station on Signy Island, in the South Orkney Islands. He has boundless energy, can do anything that needs to be done, and sees hazards before they become hazards. Raymond Priestley, a geologist on both Ernest Shackleton's 1907–9 *Nimrod* expedition and Robert Falcon Scott's ill-fated 1910–12 *Terra Nova* expedition, reflected on the giants of Antarctic exploration with these words: "For scientific discovery give me Scott; for speed and efficiency of travel give me Amundsen, but when you are in a hopeless situation, when you are seeing no way out, get down on your knees and pray for Shackleton." If today I were in dire circumstances and saw no way out, I'd get down on my knees and pray for Russ Manning.

Kim Crosbie—"the wee Scottish lassie," as she is known to friends—did her Ph.D. dissertation research on Cuverville Island, along the Antarctic Peninsula, and later parlayed this experience into a job as an expedition leader with some of the tour ships. Small in stature but not in leadership, Kim could drag Zodiacs ashore in icy chest-high surf and be ready to lead hardy hikers up to the top of Cuverville through waist-deep snow. On one cruise, most of her Zodiac drivers happened to be women, who dubbed themselves the GODS, the "Girls Only Driving Squadron." Kim is now involved in the management of tourism in the Antarctic through the International Association of Antarctica Tour Operators (IAATO), and the co-author of *A Visitor's Guide to South Georgia*.

T. H. (Tim) Baughman is a professor of history at the University of Central Oklahoma. As a graduate student at Ohio State University he joined an expedition to Marie Byrd Land, in Antarctica, as the token humanist, to provide some levity for the serious scientists at work in the Antarctic. Tim, as an eminent Antarctic historian with several scholarly books to his credit,[9] lectures about Antarctic history aboard cruise ships. In the ship's lecture theater he is a master storyteller, leaving audiences informed, spellbound, out of breath, with tears in their eyes. Ashore, after a dozen or more seasons in the Antarctic, he has finally learned to identify penguins.

9. Among T. H. Baughman's books are *Before the Heroes Came: Antarctica in the 1890s* (Lincoln: University of Nebraska Press, 1994); *Ice: The Antarctic Diary of Charles F. Passel* (Lubbock: Texas Tech University Press, 1995); and *Pilgrims on the Ice: Robert Falcon Scott's First Antarctic Expedition* (Lincoln: University of Nebraska Press, 1999). He has also published a short popular biography, *Shackleton of the Antarctic* (Lincoln, University of Nebraska Press, 2009).

ARCTIC TOURISM

Many of the same features that draw tourists to the Antarctic entice them to the Arctic as well. The Svalbard Archipelago, including the large island of Spitsbergen, sits between Norway and Greenland, well north of the Arctic Circle. Spitsbergen is easily accessible by both sea and air, and offers excursions to glaciers and rich wildlife viewing, including reindeer, walrus, arctic fox, polar bear, and a great variety of seabirds. The five thousand or so polar bears on Spitsbergen outnumber the human population two to one, and add a new requirement to the usual outfitting of tourist groups—a high-powered rifle in the hands of a well-trained guide.

Greenland itself is a miniature Antarctica, a landmass extending from 60° to 82° north, more than 1,500 miles south to north, and around 700 miles across. It is covered nearly entirely with a mile-thick sheet of ice, two miles at the thickest point—a volume of ice about one tenth that of Antarctica. A seven-mile-high glimpse of this frozen world can be had on flights from Europe to North America—the westward flight path usually passes close to or over the southern tip of Greenland, and on a cloudless day offers window-seat passengers an exquisite view of ice, rock, and water. The surrounding sea appears as a fabric of blue with tiny white polka dots—but they are not polka dots; they are icebergs that have spilled off Greenland, into the sea. And a closer look shows that the icebergs are not randomly adrift, but are arrayed in huge gyres tens of miles across—giant, slowly swirling eddies on the fringes of the northward-bound Gulf Stream.

But an overflight is not real tourism—it just whets the appetite for close-up encounters with the polar ice. That takes place at the surface. Small ships with tourists venture into the Davis Strait and Baffin Bay between Canada and Greenland for iceberg viewing and fjord cruising along the west coast of Greenland, following the eastern entry into the

Northwest Passage. And overland excursions are possible in northern Norway, Sweden, and Finland, including the opportunity to stay in the Ice Hotel (yes, a hotel carved entirely in ice) in the village of Jukkasjärvi, in Swedish Lapland, well north of the Arctic Circle.

The economic strains that followed the breakup of the former Soviet Union in 1989 forced the Russian fleet of icebreakers and polar research vessels to find other sources of revenue to support operations and maintenance. These ships entered the tourist trade in the polar regions, with several small research ships now regulars in providing tourism to the Antarctic. But in the Arctic, the big draw is the North Pole, and only massive icebreakers can be counted upon to grind a path to the pole through the Arctic sea ice.

The departure point for polar trips is commonly Murmansk, in the far northwest of Russia, a year-round ice-free port situated well north of the Arctic Circle, but warmed by wisps of the Atlantic Gulf Stream that wrap around Scandinavia into the Russian Arctic. Murmansk lies about 1,500 miles from the North Pole; from there, it takes the better part of a week to reach the pole by sea. The remote Franz Josef Islands mark the halfway point, and offer a rich array of polar wildlife, as well as a piece of the history of polar exploration. Norwegian explorers Fridtjof Nansen and Fredrik Hjalmar Johansen wintered there in 1896–97, following their unsuccessful attempt to reach the North Pole.[10]

The route from Franz Josef Land to the North Pole is a hard slog, but it is the kind of work that big icebreakers are built for. Sea ice ten to twenty feet thick forms a solid collar around the pole, through which a channel must be opened. One of the veteran Arctic icebreakers is the Russian ship *Yamal*, a nuclear-powered behemoth of some twenty-three thousand tons based in Murmansk. Icebreakers do not wedge ice apart with a sturdy knife-edge bow; they ride up onto the ice with a rounded

10. The Nansen expedition to reach the North Pole relied on an untested concept: drifting over the pole while locked in moving sea ice. It is described in more detail in chapter 2.

hull and break it beneath them through their sheer mass. It is a very noisy process, repeated time and time again around the clock, as the ship inches to the pole. It is not a quiet, peaceful, serene approach of a ship slicing silently through the sea, but rather a continuous and audible application of industrial-strength brute force. Two to three days beyond Franz Josef Land, *Yamal* arrives at 90° north. The passengers clamber down on the ice, form a circle around the pole for an arrival "ceremony," and then have a picnic on the ice. But the sea ice platform for the picnic table is proving less reliable—in August of 2000, *Yamal* arrived at the pole to discover only open sea.

POLAR PERILS

Neither the Arctic nor the Antarctic is a forgiving environment, a reality well known or quickly learned by the early explorers. Already there have been several mishaps that should raise the cautionary flag for polar tourism. In 1977, Air New Zealand began flyovers to Antarctica, a long journey back and forth from New Zealand for a few hours of in-flight viewing of the Antarctic landscape. This particular type of tourism came to an abrupt end in 1979 when one planeload of tourists crashed into Mount Erebus near New Zealand's Scott Station in the Ross Sea region. All 257 people aboard the aircraft perished.

On January 28, 1989, the Argentine supply vessel *Bahía Paraíso* struck submerged rocks and ripped her hull open shortly after leaving Palmer Station, a small U.S. research base on the Antarctic Peninsula. All the crew and tourists aboard took to lifeboats, and shortly thereafter were back at Palmer. The maximum capacity of Palmer is around forty people, so the influx of an extra two hundred placed substantial stress on the Palmer facilities. Two nearby tourist vessels, *Explorer* (the Little Red Ship) and *Illyria*, diverted to Palmer, picked up the survivors, and carried them northward to a Chilean base on King George Island, from which they were flown back to Argentina. Tides lifted *Bahía Paraíso* off the fatal rock, from which she drifted across the bay and rolled over in shallow water. Her rusting hulk can still be easily seen by passing ships today.

Probably the most visited destination along the Antarctic Peninsula is Deception Island, a heavily glaciated active volcano. Deception Island has a big interior caldera, analogous to Crater Lake in Oregon, but flooded with seawater because of a narrow breach in the wall of the volcano that connects the open sea with the sheltered interior caldera. The caldera has provided safe haven for mariners since at least the middle of the nineteenth century, and was the site of an extensive whaling operation in the early twentieth century. The breach through the wall of the volcano is visible from only one azimuth—from all other approaches Deception appears to be just another island in the South Shetland archipelago, hence the name Deception.

The passage from the open sea into the interior anchorage requires very careful piloting through the breach, because a big shallowly submerged rock ledge obstructs the middle of the channel. That obstacle restricts entry and egress of ships to an even narrower but deeper route close to the wall of the channel. The rock in the middle is perhaps the "best-known rock in Antarctica," because over the 150 years or so that mariners have known of the narrow channel to the sheltered interior of Deception Island, the rock ledge in the center of the passage has been very well mapped and charted. Hundreds if not thousands of passages by explorers, whalers, scientific survey ships, and tourist vessels have made this dangerous spot abundantly clear. And for those who need a visual rather than cartographic reminder, there is a rusted hull of a broken ship just inside the caldera that offers mute testimony to the perils of ignoring this navigational hazard. Nevertheless, on January 30, 2007, the Norwegian cruise ship *Nordkapp* damaged her hull on the rock upon exiting the caldera, and was forced to retreat into the anchorage and seek emergency assistance from a British Antarctic Survey research vessel to evacuate the 280 passengers and some 50 nonessential crew members.

Ironically, the Little Red Ship *Explorer*, the pioneer of adventure tourism, ultimately went to rest at the bottom of the sea. In late 2007, in the Bransfield Strait, between the South Shetland Islands and the Antarctic Peninsula, *Explorer* hit ice that opened a ten-foot split in her hull, and began to take on water. All passengers and crew boarded lifeboats, and were rescued without loss of life, by the *Nordnorge*, another Norwegian cruise ship operating in Antarctic waters. *Explorer* had performed similar emergency duty for those taken off the sinking *Bahía Paraíso* two decades earlier. Within hours, *Explorer* rolled over and slipped beneath the surface. To those of us who had spent many happy days aboard her, it was a melancholy moment. Symbolically (but probably not environmentally) it seems a better fate for *Explorer* to rest on the ocean floor near Antarctica than to be ignominiously cut up for scrap in a Singapore shipyard. The official inquiry into this accident attributed the sinking in part to excessive speed while traversing an iceberg field.

The accidents continue. In early December of 2008, the Argentine cruise ship *Ushuaia* ran aground near Wilhelmina Bay, on the west side of the Antarctic Peninsula, and had to evacuate more than eighty tourists. Most of the crew remained aboard, trying to contain a fuel spill that surrounded the ship to a distance of a half mile. And in early 2009, the *Ocean Nova* ran aground in Marguerite Bay. All sixty-five passengers were evacuated to another cruise ship in the vicinity, which returned them to Argentina. The ship's hull was dented but not pierced.

IS TOURISM RUINING ANTARCTICA?

What are the consequences of so many ships and tourists coming to Antarctica? As a trip to Antarctica draws to a close, and the ship heads north for the return crossing of the Drake Passage to South America, many visitors to Antarctica become pensive. The impact of the continent on visitors is often partly spiritual; they have just experienced something only a privileged few can ever hope for. Their impressions always include a sense of how vast, how unoccupied, how unsullied, how pristine Antarctica is. They see it as a frozen outpost of creation without the ubiquitous overprint of humanity seen on all the other continents. And most visitors want it to stay that way, although one still hears the occasional inquiry as to when there will be hotels and casinos in Antarctica.

Inevitably I am asked, "Are we damaging Antarctica when we come here? Are we bothering the penguins and seals by inserting ourselves, however fleetingly, into their natural world?" The question is a thoughtful one, and as tourism in Antarctica has developed, so has the research examining the impact of relatively large numbers of visitors on the terrain and wildlife of the frozen continent. The International Association of Antarctica Tour Operators has developed behavioral guidelines for visitors, addressing wildlife viewing, avoidance of fragile moss-covered areas, safeguards to prevent the introduction of non-native plants and microbes, and other issues, such as noise, littering, graffiti, and removal of natural specimens and historical artifacts. The adage "Take nothing but pictures, leave nothing but footprints" is too lenient for the Antarctic—a footprint on a pad of moss may remain there for decades, so slow is the pace of regeneration in the polar environment.

Not surprisingly, there have been some adverse environmental consequences associated with the accidents involving tourist vessels. The grounding of the *Bahía Paraíso* in 1989 released between 160,000 and 180,000 gallons of fuel that within a few days produced an oil slick that spread over twelve square miles. Limpets and algal mats in the intertidal zone were significantly impacted, seabirds less so, and fish and marine mammals negligibly.[11] In the cold environment of the Antarctic Peninsula, microbial degradation of the fuel spill was slow.

11. M. C. Kennicutt, "Oil Spillage in Antarctica: Initial Report of the National Science Foundation–Sponsored Quick Response Team on the Grounding of the *Bahía Paraíso*," *Environmental Science & Technology* 24 (1990): 620–24.

Research into the impacts of tourism on wildlife generally shows, however, that well-behaved tourists are more curiosities than disturbances to wildlife. Experiments on islands with separated penguin rookeries, where one breeding area is exposed to tourism and the other is sheltered from it, indicate few if any touristic impacts on breeding success.[12] Tourists can go home comforted in the knowledge that they have been good stewards while in the blue-and-white Garden of Eden.

But that is only part of the answer about whether they have inflicted damage on the Antarctic landscape and ecosystems. When I am asked that question, I tell visitors that it is not what they do during their two weeks in Antarctica that damages the white continent. No, it is what we all do at home the other fifty weeks of the year that is damaging Antarctica. It is our intensive use of fossil carbon-based energy to fuel a seemingly insatiable consumptive lifestyle that is warming the planet and causing irreversible changes in Antarctica.

Globalization is more than telecommunications and an integrated worldwide economy. Earth's atmosphere has always been globalized—when we deliver climate-changing greenhouse gases to the atmosphere in the Northern Hemisphere, it is not long before the effects of that atmospheric pollution are communicated to the rest of the world. The Antarctica that tourists see today is already different from the Antarctica encountered by nineteenth-century explorers, or even that seen by earlier tourists only two decades ago,[13] and more changes are yet to come.

---

12. W. R. Fraser and D. L. Patterson, "Human Disturbance and Long-term Changes in Adélie Penguin Populations: A Natural Experiment at Palmer Station, Antarctic Peninsula," in B. Battaglia, J. Valencia, and D.W.H. Walton, eds., *Antarctic Communities: Species, Structure and Survival*, Scientific Committee for Antarctic Research (SCAR), Sixth Biological Symposium (New York: Cambridge University Press, 1997), 445–52.

13. The changes taking place in Antarctica, indeed all around the world, are laid out and discussed more fully in later chapters.

CHAPTER 2

ICE AND LIFE: ON EARTH AND BEYOND

I long to see those icebergs vast,
With heads all crowned with snow;
Whose green roots sleep in the awful deep,
Two hundred fathoms low.

—WILLIAM HOWITT (1792–1879)
"The Northern Seas"

Ice is a common material with uncommon properties: it can flow downhill like a river, carve rock like a chisel, reflect sunlight like a mirror, and float on water like a cork. On a human scale it is a platform for wintertime fishermen, an arena for combative hockey players, a stage for graceful figure skaters, and an integral component of Scotch on the rocks. George Washington threaded his way through it in the historic Christmas crossing of the Delaware. The *Titanic* sank after colliding with it on her maiden voyage in 1912. It is a central element of global catastrophe in Kurt Vonnegut's *Cat's Cradle*. And ice is cold to the touch.

The chill of ice is not only a welcome therapy for strained muscles but also a property widely taken advantage of a century ago in domestic refrigeration—the icebox.

THE FIRST REFRIGERANT

I was born in 1936, and as a boy growing up in the forties, I remember well my mother reminding me to "put the milk back in the icebox." Of course, even way back then, we didn't have an icebox. We had a modern refrigerator that took advantage of an electric compressor and the thermodynamic principle that compressed gases cool upon expansion. But as late as the 1920s, ice was still widely used as a refrigerant, and the household cooling chamber was called the "icebox." Both my father and mother had grown up in homes with iceboxes.

Ice for iceboxes was an important commodity—it was cut from rivers and lakes in the winter, and distributed year-round in urban areas. It was a renewable resource, at least in years when nature cooperated. The annual ups and downs of the ice supply were discussed in agricultural terms—an 1886 newspaper headline declared "The Hudson River Ice Crop; A Full Harvest and No Fear of a Summer Famine."[1] When the crop did fail, New York imported ice from Maine by ship or, when really desperate, from Canada or Norway. And foreshadowing modern environmental issues, there was concern about the quality of the harvested ice, stemming mainly from domestic sewage and industrial discharges upstream.

The distribution of ice was also a big industry. Deliverymen carrying blocks of ice with tongs deposited the ice in special street-side chutes in buildings and houses. In 1882, New York City had up to a thousand delivery wagons, pulled by two thousand horses, delivering a million tons

1. *New York Times,* February 13, 1886.

of ice annually. And that level of consumption required production of two million tons of ice, because of shrinkage due to melting over the year.

Modern refrigeration began in the late 1920s with the invention of safe synthetic refrigerants and the widespread availability of electricity. The new refrigerants, members of a chemical family known as chlorofluorocarbons (CFCs), revolutionized the cooling industry and ended the era of iceboxes. Toward the end of the twentieth century, however, the CFCs themselves were removed from production worldwide. The reason? Five decades after their introduction, atmospheric scientists discovered an unintended consequence of the CFCs—these molecules played a central role in the destruction of stratospheric ozone and the creation of the "ozone hole" over Antarctica. Stratospheric ozone serves as a filter that prevents much of the solar ultraviolet radiation from reaching Earth's surface. It is nature's own planetary-scale sunscreen, and when this ozone filter is thinned or breached, we have to lather on the human-made stuff, cover up, and wear sunglasses.

The 1995 Nobel Prize in chemistry was awarded to Sherwood Rowland, Paul Crutzen, and Mario Molina for their discovery of how the CFCs depleted the stratospheric ozone. It was the first and so far the only Nobel Prize in chemistry to be awarded for research in environmental chemistry. It is ironic that much of our knowledge about ozone—its regional and seasonal variation over the globe, and its distribution with height above Earth's surface—was the result of much earlier research in the 1920s, at the very same time when the first seeds of ozone destruction were being sown. In retrospect, the CFCs destroyed both the icebox industry and stratospheric ozone.

NATURE'S REFRIGERATOR

Ice has also preserved the recent evolutionary history of the woolly mammoths, close relatives of modern elephants and one of the most widely recognized examples of the vertebrate fauna of the last ice age. Woolly mammoths frozen in the Siberian permafrost were reported by some of the earliest European explorers of the vast tundra environments of northern Asia. In 2007, a Siberian reindeer herder discovered a complete infant female woolly mammoth, fully articulated, with skin and internal organs intact. Dan Fisher, my colleague at the University of Michigan Museum of Paleontology, was part of the six-person international team that first examined the little mammoth just a few weeks after her discovery.[2] Studies now under way on the infant's not-yet-erupted tusks and molars will reveal the season of her birth, the richness of her mother's milk, variations in air temperature over her short life, and possibly the cause of her death.

In this particular case the soft tissues are also available for study, including a generous layer of subcutaneous fat that indicates the animal was well nourished. The scientific importance of such finds is not just to know a little bit more about the life of one baby mammoth, but rather to validate the methods by which paleontologists infer conditions of life from the structure and composition of tusks and molars. In this way, a few well-preserved specimens, recovered from the frozen archives of the North, yield keys for answering much larger questions about the history of life and climate on Earth.

Similarly, the 1991 discovery of the frozen 5,300-year-old "Iceman" at the terminus of a melting Alpine glacier has provided rare insight into the lifestyle of this not-too-distant relative of modern Europeans. He was fully clothed—attired in leather, a bearskin cap, and a woven grass cloak. His lower body skeletal elements indicated that he was a hiker. His intestines revealed a last meal of deer meat, wheat bran, and fruit. Elevated concentrations of copper in his hair suggest that he practiced copper smelting, and forensic evidence indicates that he met a violent

2. Tom Mueller writes about the baby mammoth and Dan Fisher in the article "Ice Baby," which appeared in *National Geographic* magazine in May 2009.

death. It is only because his body, and that of the baby mammoth, were preserved by ice that we can reconstruct such a complete picture of their lives.

THE SPECIAL PROPERTIES OF ICE

What is ice? Simply solid H_2O—or, as it is more commonly known, just plain frozen water.[3] Two hydrogen atoms are chemically bonded with one oxygen atom—a simple molecule that typically aggregates in a hexagonal crystal lattice, most easily seen in individual snowflakes. One very unusual property is acquired when liquid water freezes to form solid ice. Most liquids contract in volume when they solidify, but when water freezes, it expands. The same mass of H_2O occupies a greater volume in solid form than in liquid form, and accordingly the solid has a lower density than the liquid. This property enables ice to float in water, a phenomenon seen at scales ranging from ice cubes in a beverage glass to huge icebergs afloat in the oceans. No other common substance becomes less dense as it solidifies.

Everyone has heard the saying "It's just the tip of the iceberg," generally signifying that there is much more to something than meets the eye. But how much more ice is there beneath the surface of the water, compared to what we see above the surface? The answer is usually given as some multiple of the height of the visible ice—eight or nine or ten times—a number related to the fact that ice is only nine tenths as dense as water. But the proper comparison is between the mass above the water and that below. That, in turn, depends on the shape of the berg—a pyramidal berg and a cube of the same volume would have different root geometries.

3. Some other frozen substances are also called ices, with perhaps the best known being frozen carbon dioxide, which is also called dry ice.

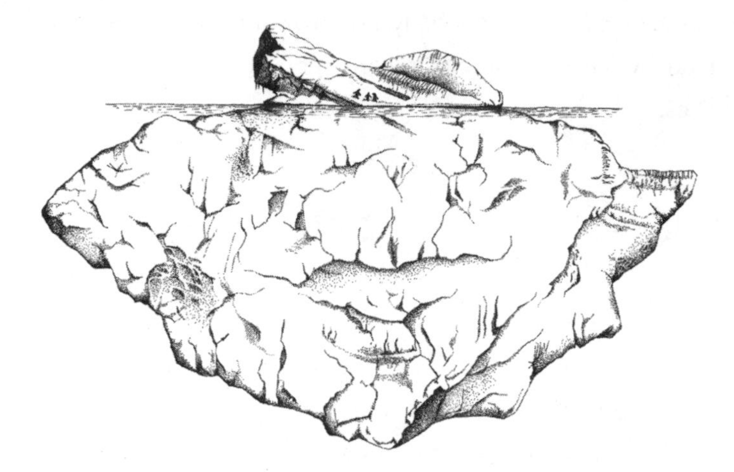

An iceberg afloat, with the "tip" above water and the "root" below the surface

Another fascinating property of ice is its propensity to flow downhill as a glacier. The ability to flow is something we generally think of as a property of fluids, and thus we are led into a seeming paradox: A flowing solid? The resolution of this paradox lies in the fact that most solids, when they are close to their melting temperature, lose their rigidity and become soft. A stick of butter just out of the refrigerator is stiff and brittle, but that same stick left on the kitchen table will warm and become soft and ductile, even though still solid. Similarly, a candle is stiff, except near the flame, where it softens considerably. So, too, with ice—even though ice seems very cold to the touch, in reality it is very close to its melting point of 32° Fahrenheit, and so weak and ductile that, given sufficient time, it will spill off hills and plateaus and flow downhill like a river, even in its solid state.

Weakness in solids is not related only to temperature—it is also a question of time. Silly Putty, that favorite pink claylike material that delights children and adults, exhibits different behaviors depending on how fast it is stressed. It will snap like a twig if yanked apart, and bounce like a rubber ball if dropped. But it can be shaped like modeling clay if kneaded slowly. Similarly, rocks that will flake or fracture when hit with a hammer will bend and flow in response to steady geological forces applied over millions of years. The Appalachian Mountains in the eastern United States appear as a wide zone of valleys and ridges—giant wrinkles in Earth's rocky crust that formed in the slowly closing vise of a collision between Africa and North America.

Ice does not really need a hill to flow down—it can spread sideways under its own weight. As snow accumulates in a region, the older snow is slowly compressed by the weight of the newer snow above it, and eventually the pressure recrystallizes the snow into ice. When enough ice accumulates and reaches a critical thickness, it will begin to squeeze the deeper ice outward, much like what happens when pancake batter is poured onto a griddle. Even though the batter is poured in the center of the griddle, it doesn't stay there—it spreads radially outward into a sheet, eventually to cook into a pancake. So it is with ice on Earth; thick accumulations of ice spread outward to form a continental-scale ice sheet thousands of miles across but only a mile or two thick.

The spread of ice over the continents during an ice age and the flow of ice from mountains downward toward the sea have profoundly shaped the landscape of Earth. Like a massive bulldozer, flowing ice erodes and transports the material over which it moves; soil, rock, and vegetation all yield easily to its force. The beautiful deep fjords incised into the coasts of Norway, Alaska, Chile, and New Zealand are all products of severe glacial erosion, an aesthetic inheritance from the recent ice ages. And when glacial ice eventually melts, the debris of glaciation is simply dropped in place, forming a blanket of glacial deposits that becomes yet another layer recording a dramatic episode in Earth's history. Some of these deposits, rich in organic matter, comprise today's bountiful soils that yield food for the people of the world.

Another important property of ice is its reflectivity. Everyone who participates in outdoor recreation in the winter—skiers, skaters, hikers, snowmobilers—knows or painfully learns the perils of sunburn from sunlight reflected upward from snow and ice. All the surfaces that make up the face of the Earth—the soils, the rocks, the vegetation, the oceans, and the ice reflect some sunlight back to space, but none does a better job than snow and ice. The property of a surface that expresses how much incoming sunlight is reflected away is called its albedo, the Latin word for whiteness. A surface with an albedo of 25 percent means that one quarter of the light that falls on that surface is reflected back to space. Not surprisingly, surfaces such as black rocks, rich dark topsoil, and dark green vegetation reflect much less sunlight, and accordingly have smaller albedos—they are not very "white."

The reflectivity of ice, particularly as compared with water, has played an interesting role in navigating through polar waters. On the fringes of both the Arctic and Southern oceans, the breakup of sea ice in the spring and summer produces areas of sea ice and of open water. The challenge facing a navigator aboard a ship is to find the passages of open water between the sheets of sea ice. That is where knowledge of albedos can be a big help.

White sea ice is a much better reflector of sunlight than dark seawater or ice-free land. When there are clouds overhead, the more abundant light reflected from the ice illuminates the base of the clouds with diffuse light, much like the nighttime lights of a city cast a glow in the sky that can be seen many miles away. To the polar navigator, the bright regions of distant clouds signify ice below, and the darker regions signify open water. Early navigators had a phrase for this: "watch the clouds to find the water." This phenomenon is colloquially termed "ice blink" or "water sky," and it is referred to frequently in the navigational logs and diaries of polar explorers. It is a phenomenon equally well known in the cultures of the Arctic's native peoples.

As Captain Cook sailed beyond 71° south in January of 1774, his farthest point south in all of his three voyages of discovery, he knew from ice blink that he was close to the ice margin, well before he actually saw the ice:

> We perceived the clouds over the horizon to the south to be of an unusual snow white brightness, which we knew announced our approach to field ice . . . the southern half of our horizon was illuminated by the rays of light which were reflected from the ice to a considerable height.[4]

On a much larger scale, the high albedos of ice and snow influence Earth's climate substantially, because the polar ice caps reflect away much of the sunlight that illuminates the high latitudes. With the rapid summertime loss of Arctic Ocean sea ice that is now taking place, the albedo of the Arctic Ocean is also changing. The balance between white ice and dark ocean water is tilting toward darkness, and the albedo is gradually falling, leading to less sunshine being reflected away and more being absorbed. In climate terms, this imbalance means the Arctic is becoming warmer because it is retaining more solar energy.

Ice interacts with light in ways other than simple reflection. Reflected light usually looks much the same as incoming light. The act of reflection plays no favorites in terms of colors—there is no selective enhancement of any of the colors that make up visible sunlight. But this is not the case for the light that is transmitted through the ice. The molecular bonding in the H_2O molecule, the crystalline structure of ice, and impurities such as tiny bubbles, together act like a filter, removing the red and yellow end of the visible spectrum but letting the blues pass through relatively unimpeded, thus giving extraordinary shades of blue to deep snow and glacial ice.

NATURALLY OCCURRING ICE is widespread on Earth, but it is not found everywhere. In fact, as ubiquitous and familiar as ice may seem, there are places in the world where ice once was, and perhaps still is, a curious and unfamiliar material. In *One Hundred*

4. This extract from Cook's journal is from Hough, *Captain James Cook*, 236–37.

Years of Solitude, set in his native Colombia, Gabriel García Márquez describes such an encounter:

> . . . there was a giant with a hairy torso and a shaved head, with a copper ring in his nose and heavy iron chain on his ankle, watching over a pirate chest. When it was opened by the giant, the chest gave off a glacial exhalation. Inside there was only an enormous, transparent block with infinite internal needles in which the light of the sunset was broken up into colored stars. . . . José Arcadio Buendía ventured a murmur:
>
> "It's the largest diamond in the world."
>
> "No," the gypsy countered. "It's ice."
>
> José Arcadio Buendía, without understanding, stretched out his hand toward the cake, but the giant moved it away. "Five reales more to touch it," he said.
>
> José Arcadio Buendía paid them and put his hand on the ice and held it there for several minutes as his heart filled with fear and jubilation at the contact with the mystery . . . with his hand on the cake, as if giving testimony on the holy scriptures, he exclaimed: "This is the great invention of our time."[5]

Where on Earth does ice occur naturally? The simple answer: wherever there is water available to freeze and it is cold enough to freeze it. Let's have a look first at what controls Earth's temperature at different locations. Over the surface of our planet there is a wide variation in temperature related to the angle at which the Sun's rays strike Earth. At the equator, the rays are nearly perpendicular to the surface, and near the poles they just graze the surface. Effectively that means that the farther away one is from the equator, the less solar heat is available for warm-

---

5. Gabriel García Márquez, *One Hundred Years of Solitude*, trans. Gregory Rabassa (New York: Harper & Row, 1970).

ing the planetary surface. This establishes a general pattern of gradually diminishing temperature toward the poles—with the poles averaging (on an annual basis) about 90 Fahrenheit degrees colder than equatorial regions, or a drop of roughly 1 Fahrenheit degree for every 70 miles one travels away from the equator toward the poles. It should be no surprise that the poles are the principal domains of ice on Earth, both on the continent of Antarctica and over the Arctic Ocean and Greenland.

There is also a very noticeable decrease in temperature as one goes up into the atmosphere, much more so than heading toward either pole. Over the entire distance from equator to pole, about 6,250 miles, the temperature drops by about 90 Fahrenheit degrees, but that drop in temperature can be experienced also by going only five miles up into the atmosphere. When the pilot of a passenger aircraft announces that the plane has reached its cruising altitude of 35,000 feet and the outside air temperature is −55° Fahrenheit, that frigid temperature is the result of the airplane's climbing up into the atmosphere.

Ice also forms in the atmosphere when water vapor is carried sufficiently far upward, giving rise to snow during winter months and occasional hail during severe summer storms. Hail usually falls in the form of spherules, ranging from pebble to baseball size, and has the potential to be both damaging and dangerous. It can flatten crops in a matter of minutes, leave cars with a pitted complexion, and injure people and animals unable to find shelter.

Some tiny ice crystals also form high in the frigid polar stratosphere, where they provide catalytic surfaces for separating chlorine from man-made chlorofluorocarbons, the CFCs that revolutionized the refrigerator industry. The freed chlorine in turn destroys stratospheric ozone each spring, thus opening the seasonal ozone hole over Antarctica. The consequence of this ozone depletion is increased ultraviolet radiation reaching Earth's surface, with attendant radiation damage to the biosphere. Humans in particular are vulnerable to skin cancers, cataracts, and immune system impairment.

Temperature also changes going down from the surface, into the rocks of Earth's crust. But in the downward direction, the temperature goes up, an indication that the interior of Earth is losing heat to the surface. The rate at which the temperature increases downward is about five times the rate that it cools upward in the atmosphere. Five miles up in the atmosphere it is 90 Fahrenheit degrees cooler, but at a depth of only one mile below the surface, it is 90 Fahrenheit degrees warmer than at the surface. That is the reason that chilled air must be pumped into deep mines, to make it possible for miners to work. At much greater depths the temperature in places reaches the melting point of rock and produces volcanic magma.

But in the polar latitudes, where the average surface temperature on land is well below the freezing point over long periods of time, water in the subsurface will freeze despite the fact that the temperature increases with depth. This freezing gives rise to perennially frozen ground known as permafrost. Eventually, of course, one reaches a depth where heat coming from Earth's deeper interior has warmed the rocks enough to prevent permafrost from forming. That depth defines the lower margin of permafrost. Permafrost has been observed to almost a mile below the surface in northern Siberia, but in areas where the average annual temperature is only slightly below freezing, the permafrost is much thinner. About one fifth of Earth's land area, mostly in the northern reaches of Asia, North America, and Europe, is characterized by permafrost today.

Ice also forms in sediment beneath the seafloor, in a crystal structure that is in the shape of a not-quite-spherical cage. The depth at which this form of ice occurs is relatively shallow, some five hundred feet or so below the seafloor, within the sedimentary deposits. This particular type of ice is widespread on the continental shelves and, on rare occasions, in deep lakes on the continents, such as in Siberia's Lake Baikal. What makes this ocean floor ice of special interest is its ability to trap methane—natural gas—inside the molecular cage. Drilling into the shelf sediments has retrieved samples of this gas-bearing ice from a great many sites around the world. When ignited, the samples put on quite a show—to see chunks of ice aflame is completely intuition-defying. Methane, of course, is an important source of energy in the industrialized world, but it is also a potent greenhouse gas when it resides in the atmosphere. The release of methane from its icy subsurface cage is therefore a worry in the context of climate change.

We typically think of water becoming ice in terms of temperature alone, but pressure also plays a role. The pressure under which ice forms at the surface of Earth is more or less uniform—it is the pressure caused by the weight of the atmosphere lying above the surface. Informally, this amount of pressure has been called "one atmosphere" or "one bar." Under this amount of pressure, freshwater will freeze when the temperature drops below 32° Fahrenheit.

Even the pressure of the atmosphere is somewhat variable at Earth's surface. In the eye of a hurricane the atmospheric pressure is lower than average, and on a cold, clear, sunny winter day the pressure is above average. Even though this variability of pressure from place to place is small—typically only a few percent from the average—the highs and lows play a big role in the pattern of atmospheric circulation and our daily weather.

Bigger changes in pressure occur with a change in elevation. Along with the temperature in the atmosphere, as one goes up, the pressure goes down. At sea level, the entire atmosphere is above you, but at the top of Mount Everest, only about half the mass of the atmosphere presses on you, exerting only a half atmosphere of pressure. But even at elevations far below the summit of Mount Everest, the thinning of the atmosphere becomes apparent—where there is less atmosphere, there is less oxygen to breathe, and even simple tasks become arduous. And when the atmospheric pressure is less, water boils at a lower temperature, thus requiring longer to cook rehydrated meals and making water purification less reliable—facts that every experienced mountain camper knows well. Lower atmospheric pressure also shifts the freezing temperature of water very slightly upward.

Let us pause to reflect on what a very special thermal environment the Earth's surface provides us, sandwiched between extreme heat downward and ice upward, in just a few miles in either direction. Other life-forms, mostly microbes, can eke out an existence in these extreme thermal environments, but we humans, pansies in terms of our preferred temperature, stick pretty close to sea level. Our comfort zone is a very thin envelope of oxygen-rich atmosphere just above Earth's surface—a Goldilocks environment where it is not too hot and not too cold. When we turn to look at the possibilities of life elsewhere in the solar system later in this chapter, we will see that Earth itself is sometimes called the Goldilocks planet, but for a different reason.

ASIDE FROM TEMPERATURE, the other part of the recipe for ice is the availability of water. H_2O takes up residence in many places on Earth, in places appropriately called reservoirs. The principal reservoirs are the oceans, the polar ice sheets, lakes and rivers, underground water, the atmosphere, and the biosphere. H_2O moves from one reservoir to another by evaporation, precipitation, biological transpiration, and the downhill flow of water and ice. Of the reservoirs, clearly the worldwide ocean is the largest by far, holding almost 96 percent of Earth's H_2O. After the oceans, the next biggest reservoir is ice, with a little more than 3 percent of the global H_2O. That leaves only 1 percent for everything else—all of the surface, subsurface, atmospheric, and biologic water combined. The volume of water in lakes, rivers, groundwater, atmospheric water vapor, and vegetation is small compared to that in ice, and very small compared to the volume of ocean water.

On land, the presence or absence of water depends strongly on the transport of water vapor from the sea to the land by atmospheric circulation. In some places—Earth's deserts—the atmosphere delivers almost no water. Other areas receive abundant rain and snow, and have become places of human settlement and agriculture, and tropical, temperate, and boreal forests. At low elevations outside the polar regions, the precipitation yields wetlands, lakes, and rivers, some of which experience seasonal freezing, and at higher elevations, the precipitation produces snowpack and glacial ice. In the polar regions, snow and ice dominate at every elevation.

SEA ICE

In the oceans one need not search for water—it is very apparent and readily available for freezing when the temperature is sufficiently cold. As one gets closer to the poles, the temperature at sea level approaches the freezing mark, and the ocean water itself will freeze into sea ice. Seawater, because of its salt content, freezes at a slightly lower temperature (28.8° Fahrenheit) than freshwater. Annual sea ice—ice that forms and breaks up each year at high latitudes—is about three feet thick. The Antarctic continent effectively doubles in area—to a size exceeded only by Asia and Africa—as the bordering Southern Ocean freezes each winter.

Watching the sea freeze over is an eerie experience. It begins rather innocuously with the formation of small pancakes of ice that grow laterally until they meet other pancakes. At the common boundaries, small ridges form and intersect to form an array of polygons, a geometric tiling of the ocean surface that forms a semi-rigid mat extending as far as the eye can see. Under continued cold conditions the ice pancakes thicken, and in the process of freezing, the salt is expelled from the solidifying ice. It forms dense plumes of brine that sink from the base of the newly forming ice and descend all the way to the bottom of the ocean. And all of this happens rather quickly. An inexperienced captain can discover that in just a short time his vessel has lost the freedom of the open sea and is incarcerated by the tight grip of ice. A great fear of early sea captains exploring in the Arctic and Antarctic was the possibility of being trapped by a rapid freezing of the sea, and forced to winter over, cold and hungry, until the ice broke up the following spring.

Such concerns about sea ice figured prominently in the exploration of the Bering Strait, that narrow strip of ocean that separates Asia from Alaska. Vitus Bering, a Danish sea captain in the employ of Czar Peter the Great of Russia, was given the assignment to determine whether Asia was connected to North America. We know today that the two continents are not connected, but early in the eighteenth century, that fact of geography was still uncertain. The plan to determine the existence of a strait was straightforward: along the Arctic coast in the far east of Russia, there was a small trading post where the Kolyma River emptied into the Arctic Ocean. If one could reach this trading post by sea from the Pacific Ocean, it would prove that there was open water between Asia and North America.

Traveling close to the coast of Russia in mid-August of 1728, Bering actually sailed from the Pacific into the Arctic Ocean, but he was not certain that his course had taken him through a narrow strait because he could not see North America to the east. Bering continued north, but even in late summer the sea was showing signs of an imminent freeze. He decided not to turn west, toward the mouth of the Kolyma River, for fear of being trapped in the ice for the winter months. Bering ordered a course reversal, and the ship returned to the Pacific.

Even though never sighting North America and never reaching the Kolyma River—the position of proof—Bering felt he had gone far enough to demonstrate the existence of a strait between Asia and North America. When he returned to St. Petersburg, however, the armchair geographers felt otherwise. They criticized his judgment in choosing safety over mission, denied him a promotion, and shamed him into a second voyage. On that voyage, a shipwreck forced him to make winter camp on a remote treeless island at the very end of the Aleutian Islands arc. He and many of his crew perished over the winter; come spring, the few survivors salvaged timber from the wrecked ship, built a new, smaller ship, and sailed away to tell the tale.

SEA ICE ON THE MOVE

Near the end of the nineteenth century, many of the geographic "targets" of explorers—the source of the Nile, the Khyber Pass, Timbuktu—had been reached. But the poles of Earth, both North and South, remained unconquered. The movement of the Arctic sea ice figured prominently in Norwegian explorer Fridtjof Nansen's attempt to be the first to reach the North Pole. Nansen, already well known for his first-ever crossing of Greenland on skis in 1888, had a clever idea of how to reach the North Pole with a minimum of effort. Because Arctic sea ice is always on the move, drifting generally from Siberia toward Scandinavia and Greenland, Nansen would let his ship become frozen into the ice in the far east of the Arctic Ocean, and simply drift with the moving sea ice on its way toward Greenland and its ultimate breakup and escape into the North Atlantic. If one understood the dynamics of the drift well enough, and entered the ice at the proper location, the drifting sea ice, moving at around three to four miles per day, would carry the ship over the pole—a leisurely cruise, so to speak, without an epic struggle against nature, a triumph of cleverness over brute strength.

Nansen's ship *Fram* ("forward" in Norwegian) had been designed for such a voyage, incorporating a rounded hull that would enable the ship to be lifted upward by the pressures within and between sheets of drifting sea ice, rather than risk being crushed with a sharper, more conventional hull architecture. Nansen estimated that, once drifting with the ice, the ship would reach the pole in about a year and a half, and continue on to Norway in perhaps a similar amount of time. In two to three years the North Pole would be his, and the world would know it.

The devil, of course, was in the details—finding the right spot to enter the ice, so that the drift trajectory would carry the ship to the pole. He would have only one chance to get it right—once the *Fram* was locked into the ice there would be no opportunities for course corrections.[6] Nansen chose to enter the ice near the New Siberian Islands, a little east of the mouth of Russia's Lena River. The entry was in late September 1893, just at the onset of the Arctic winter.

Unfortunately, his entry point was not quite right. By the end of 1894, after more than a year of drifting, it was clear from astronomical observations that the ship would not pass over the pole. In fact, there was little chance it would even get beyond 85° north, some 350 miles from the pole. With the possibility of the ship achieving the pole already lost, the thought of spending yet another year or even two aboard the ice-locked drifting *Fram* before she was released to the sea was too painful for Nansen to consider.

Not one to give up the polar quest easily, he decided to leave the ship in the care of his crew and, with a single companion, Fredrik Hjalmar Johansen, set off across the ice with dogs, skis, a sledge, and kayaks, in an attempt to reach the pole. Nansen and Johansen left *Fram* in late February of 1895 and seemingly were making good progress, but by early April, observations of the position of the Sun showed that the pole was still 250 miles away. Six weeks of hard travel on the ice had brought them only 100 miles closer to their goal. An awful realization hit them: as they were struggling north, the ice was drifting south, erasing some of their daily progress toward the pole. On April 8, they recognized the futility of the attempt, and turned around.

Nansen and Johansen knew they would have little chance of finding *Fram*, because in the six weeks that had elapsed since they left the ship,

6. This is somewhat reminiscent of a story about the mid-nineteenth-century pioneer wagon trails across America. Where the California and Oregon trails diverged into two well-worn paths on the high plains of western Nebraska, there was posted a sign that read, "Choose your rut carefully—you'll be in it for the next two thousand miles."

she had continued to drift with the ice. They would have to make the homebound journey back to Norway, a distance of some 1,400 miles over the Arctic ice, entirely on their own. After many more months of adventure, hardship, and an improbable encounter with another polar adventurer, Nansen and Johansen set foot on Norwegian land in mid-August of 1896. Meanwhile, *Fram* had continued her long and slow drift across the Arctic Ocean, to break out of the ice on the very day that Nansen and Johansen reached Norway. Within a week the entire expedition had been reunited.

The drifting of sea ice also figured dramatically in the saga of Ernest Shackleton's Imperial Trans-Antarctic Expedition of 1914–16. Shackleton had already participated in two unsuccessful attempts to reach the South Pole, the *Discovery* expedition led by Robert Falcon Scott, in 1901–3, and the *Nimrod* expedition under his own leadership, in 1907–9. The first fell short by 530 miles; the second reached 88°23' south, only 97 miles from the pole. The overland segments of both expeditions turned back shy of the polar goal because of deteriorating health, diminishing provisions, and concerns that if they pressed ahead they might perish.

In 1911, Robert Falcon Scott and Roald Amundsen mounted rival expeditions to reach the South Pole. The ship that brought Amundsen to Antarctica was none other than *Fram*, the one Fridtjof Nansen had used in his unsuccessful attempt to drift over the North Pole. Amundsen and Scott chose different pathways to the pole, with Scott's slightly longer but better known. The race to the pole was won by Amundsen, by more than a month; his small team reached 90° south on December 11, 1911. Scott arrived at the pole thirty-five days later, only to endure the great disappointment of seeing the Norwegian flag atop a small tent left by Amundsen. But that disappointment was eclipsed by the death of Scott and the other four in his polar party during their return from the pole.

With the attainment of the South Pole then in Amundsen's book of laurels, and Scott's as well posthumously, Shackleton was forced to envision a different kind of exploration, a bold expedition that he hoped would ensure his own lasting prominence in Antarctic exploration. He devised a plan to cross the entire Antarctic continent, on a route from the Weddell Sea to the Ross Sea, with a stopover at the South Pole as a midway point. This expedition was to become known as the Imperial Trans-Antarctic Expedition, and the ship that carried Shackleton into the Weddell Sea was the *Endurance*.

Just as *Fram* did intentionally in the Arctic, *Endurance* did unintentionally in the Antarctic—it became locked in the ice of the Weddell Sea in January of 1915, and began a slow clockwise drift, following the current gyre in the sea beneath. The drift carried *Endurance* within some sixty miles of the ice shelf at the southern margin of the Weddell Sea, where Shackleton had planned to disembark and begin the overland trans-Antarctic trek. But the ice took them no closer—in fact, it began to carry them slowly away from the southern coast of the Weddell Sea and toward the long finger of the Antarctic Peninsula. By April, the drift exceeded two miles per day to the northwest, and the possibility of launching the overland segment of the expedition was rapidly receding. As author Alfred Lansing writes, "*Endurance* was one microscopic speck . . . embedded in nearly one million square miles of ice that was slowly being rotated by the irresistible clockwise sweep of the winds and currents of the Weddell Sea."[7]

Fram and *Endurance* had both been built by Norwegian shipwrights for service in polar regions. But *Fram* was designed specifically for a long sea ice drift—her rounded hull would accommodate horizontal pressure from the ice by popping upward as the ice closed in around it. *Endurance* was built very sturdily to push ice around, but her less rounded hull did not fare well when surrounded and squeezed by ice. Before she could be released from the ice as she moved northward toward a latitude (and season) of sea ice breakup, the ice crushed her, and she sank—*Endurance* did not endure the harshness of sea ice entrapment. The

7. Alfred Lansing, *Endurance: Shackleton's Incredible Voyage* (New York: Carroll & Graf, 1959), 37.

survival of Shackleton and his entire crew after the loss of *Endurance* is one of the moving stories of polar exploration and rescue.

ICE, WATER, AND LIFE

Water is considered an essential ingredient of life as we know it. The human body is approximately 90 percent water, so we are mostly hydrogen and oxygen. Add a little carbon and nitrogen, and you have the materials for 96 percent of our mass. Earth occupies a special location in the solar system because it is the only place in the neighborhood that has abundant liquid H_2O, that essential ingredient for life, at its surface. For this reason it is sometimes called the "Goldilocks planet"—not too close to the Sun to lose all its water, not too far to have only ice. The third rock from the Sun is just about the right temperature for the purpose of life.

But Earth almost did not qualify for hosting water at its surface. Although the amount of radiant energy the Sun delivers to Earth is adequate to make our planet hospitable to liquid H_2O, Earth reflects about 30 percent of that energy back into space, thereby rejecting a sizable fraction of the solar gift of warmth. But Earth's climate system has another important player—an atmosphere that compensates for the reflective loss. The atmosphere is 99 percent nitrogen and oxygen, but the remaining 1 percent includes very small amounts of water vapor, carbon dioxide, and methane—heat-trapping gases that block heat trying to escape from Earth in the form of long-wavelength infrared radiation. This blanketing is called the natural greenhouse effect, as opposed to the contemporary warming of the planet due to anthropogenic greenhouse gases—those of human origin—now being added to the atmosphere through the burning of fossil fuels.

The natural greenhouse effect is a characteristic of Earth's atmosphere that has been around since Earth's earliest days as a planet, whereas the anthropogenic greenhouse effect is a phenomenon of just the last few centuries of Earth history. We should be very grateful for the natural greenhouse effect, because Earth is some sixty Fahrenheit degrees warmer than it would be without it. It has made Earth the water planet, the blue planet, instead of just another white snowball in orbit around the Sun.

IS EARTH UNIQUE in the solar system as a place where life has emerged? The elements that comprise life and the rocks of our planetary host—hydrogen, oxygen, carbon, nitrogen, iron, magnesium, and silicon—are among the ten most abundant elements in the entire universe. It is clear that nature uses the most common raw materials around for both planetary architecture and life. With such materials widely available, might there be other places hospitable to life?

Of the other planets in the solar system family, Mercury and Venus are too close to the Sun to retain water, let alone to enable ice to form. But farther out in the solar system ice abounds, giving rise to fascinating possibilities of life, provided energy is available to change ice to water. Mars displays very obvious ice caps in its polar regions, two little white skullcaps sitting on the red planet. The northern cap is much larger than the southern, with a diameter some three times greater. But there is a second asymmetry that is more amazing—the two polar caps reveal a different composition, at least at their respective surfaces.

The surface of the southern cap of Mars is made up of solid carbon dioxide (CO_2), which we call "dry ice," because when it warms, the solid transforms directly into a vapor, without first passing through the liquid stage that characterizes the behavior of H_2O on Earth. Thus on Mars the transformation of solid CO_2 is a "dry" one, to a vapor, whereas the conversion of solid H_2O on Earth is a "wet" one, to a liquid, water. It is perhaps no surprise that Mars's south polar cap shows a surface composition of carbon dioxide—CO_2 is the principal component of the Martian atmosphere, and the south polar temperatures are sufficiently cold to allow solid CO_2 to condense out of the atmosphere onto the surface.

The big surprise is that the surface of the larger northern cap is not CO_2, but instead solid H_2O, the ice that we know and love on Earth. The atmosphere of Mars is much thinner than Earth's, and exerts less than 1 percent of the pressure of Earth's atmosphere. At such a low pressure, H_2O can also change from solid to vapor directly, a "dry" transformation, which can be seen indirectly as a summertime increase of water vapor over the north polar cap.

Both of the Martian polar caps probably contain substantial H_2O ice, but the H_2O in the southern cap is likely deep beneath the CO_2 ice. The reason for this compositional asymmetry is not well established, but may be related to a north pole that is warmer than the south pole, because of the presence of more summertime dust on the north polar surface, which absorbs more sunshine.

The surface of Mars away from the poles, despite having a very low atmospheric pressure on it and a temperature well below freezing today, shows many well-developed river channels that indicate that water once flowed over the surface. Layers of sedimentary rock, typically formed as deposits in lakes, rivers, and oceans on Earth, can be seen in the Martian landscape. And there is reason to suspect that some of the water that once flowed on Mars may still be there, in the form of ice just beneath the Martian surface and possibly in liquid form at greater depths. Ground-penetrating radar surveys conducted in 2008 from the Mars Reconnaissance Orbiter spacecraft returned signals indicative of massive ice deposits in the southern mid-latitudes of Mars, suggesting that there may be large glaciers buried beneath just a few feet of dust and rock debris.[8] With so much ice on Mars, the presence of water and life on the red planet is an intriguing possibility.

---

8. J. W. Holt et al., "Radar Sounding Evidence for Buried Glaciers in the Southern Mid-Latitudes of Mars," *Science* 232 (2008): 1235–38.

A TRIP TO MARS

In August of 2007, I had the special experience of watching the launch of a spacecraft to Mars called the *Phoenix* Mars Lander. On launch day at Cape Canaveral I arose well before dawn and made my way to a viewing site, on the beach a short distance from the launchpad. The tense countdown proceeded flawlessly, right to the moment of ignition. It was a thrill to see the main rocket engines ignite and roar, and, a few seconds later, to watch this huge taxi lift its small spacecraft passenger into the atmosphere, accelerate to a velocity sufficient to escape Earth's gravity, and send it on a nine-month journey to Mars. Although Mars's orbit is only 50 million miles beyond Earth's orbit, a distance half again as far as Earth is from the Sun, the actual journey of the *Phoenix* Lander covered 422 million miles, as the spacecraft had to chase Mars in its orbit, a veritable moving target.

The instruments aboard the Lander were designed to send back information about the atmosphere and soil of Mars, and in particular to seek evidence that might confirm the presence of some form of microbial life, if indeed any were present. To be sure the instruments would perform as designed, they were tested in the frigid desert environment in the Dry Valleys of Antarctica, a short helicopter ride from McMurdo Station, a U.S. base in the Ross Sea sector of the Antarctic. This cold, dry, windy setting is probably very close to the actual operating conditions on Mars.

The *Phoenix* spacecraft arrived at Mars in late May of 2008, and after a breathtaking descent through the thin Martian atmosphere, landed safely at the target landing site close to the north polar cap. I am always amazed at the precision of such operations—a NASA engineer likened it to a rocket launched in California successfully reaching its distant target—home plate in Chicago's Wrigley Field! I "watched" the landing, too, so to speak, over NASA TV's live Internet broadcast. It was an emotional moment, tying together both ends of this 269-day journey across the vastness of the solar system. *Phoenix*, never distracted from duty, announced its own arrival, waited a few minutes for the dust to settle, unfolded its solar panels to power its instruments, and, like an enthusiastic tourist, began taking pictures to send home.

The first snapshots showed "patterned ground," almost identical to earthly terrains in Siberia, Alaska, and northern Canada—low ridges of rock fragments squeezed upward into polygonal patterns by cycles of freezing and thawing. Soon *Phoenix* deployed its soil scooper, and brought into its mini-laboratory some Martian soil containing white nuggets, later confirmed to be H_2O ice. The soil also contained hints of calcium carbonate, the principal mineral of limestone, which on Earth is usually precipitated in water. Also in the soil were traces of perchlorate, an oxidizing agent that could provide nourishment for any microbial life that might be present.

But *Phoenix*'s main instrument, the one designed to detect organic compounds, proved balky, and only about half the experiments on the schedule were able to be conducted. Even though *Phoenix* continued to operate well beyond its design lifetime of three months, providing photos of dust storms and the distant faint Sun, there was no confirmed detection of an organic molecule. Of course, absence of evidence is not the same as evidence of absence, and so the possibility of some form of life on Mars remains. At the end of October of 2008, as the Sun reached the horizon, *Phoenix*'s solar panels no longer could gather enough light to power the Lander, and by mid-November it had fallen totally silent. There is little likelihood it will survive the freezing darkness of the Martian winter.

ICE (AND LIFE?) BEYOND MARS

In the more distant and frigid reaches of the solar system are the planets Jupiter, Saturn, Uranus, and Neptune—very large bodies mostly of gaseous hydrogen and helium without a solid surface of ice or rock anywhere close to their visible cloud tops. The temperature of Jupiter's cloud tops is around −250° Fahrenheit, and it only gets colder farther out. These giant planets have many small satellites orbiting about them, made up mostly of rock and ice, in varying proportions and vertically segregated—a rocky interior covered by a shell of ice hundreds of miles thick. Just as Earth's rocky crust displays faults and folds and ancient granite intrusions as evidence of a long geologic and tectonic history, so also can features in the icy crust of these distant satellites provide an archive of the processes that shaped them over time. Some of these processes may have melted ice to form water and provide an environment conducive to life.

Early in the history of the solar system there was a lot of debris flying around and colliding, creating heavily cratered surfaces such as we see on Earth's moon. If nothing else ever happened to bodies such as the moon, their surfaces today should look much as they did three to four billion years ago—heavily cratered. But if other processes are active over geologic time, surfaces can undergo alteration and modification. On Earth the ancient cratered surface has been almost totally obliterated by the process of plate tectonics, which recycles most of Earth's crust back into the planetary interior every two hundred million years or so. And what has not been recycled, what is left at the surface, is deeply eroded over time, or covered up by younger layers of sedimentary rock. In effect, Earth gets a periodic resurfacing, just as many roads—potholed and broken up during a harsh winter—receive a new layer of tarmac in the spring.

Many of the icy satellites around Jupiter and Saturn are indeed heavily cratered—the telltale signature of an ancient surface covering a dead interior. But other satellites display a very smooth surface with very few craters. Such a juvenile complexion, unblemished by the pockmarks of impacts, is an indication that the icy surfaces have experienced "resurfacing" events in their history.

How does an icy surface get rejuvenated? During a hockey game, the ice surface gets cut and scored by the players' skate blades, but between the periods of play, the surface is restored by a Zamboni, which spreads a sheet of water over the degraded ice that quickly freezes into a new smooth playing surface. So what is the great Zamboni resurfacing machine that operates in the distant reaches of the solar system? Perhaps, surprisingly, it is water coming from the interior of these planetary satellites, water that derives from melting at the base of the thick surface layer of ice. But the source of the heat that melts the ice is one that is not so familiar to Earth-dwellers—the heat derives from tidal forces exerted on the small satellites by the giants in the neighborhood, the planets Jupiter and Saturn. The nature of this heat-generating mechanism is very interesting.

TIDAL HEATING

Many people are vaguely aware of tides, but don't think much about them unless they live on the seacoast. Tides in the ocean—the rhythmic rise and fall and sloshing around of the sea—are due to interactions between the orbits and gravity fields of the Sun, moon, and Earth. What most people are unaware of is that the tidal forces move and distort not only the water on Earth, but also the solid rock of the planet, although not very much. The physics of tides tells us that the strength of the tidal force depends on the distance between the co-orbiting bodies and on how big they are. Tidal effects on Earth are small because the moon, although relatively close to Earth, is small—only 1/80 of Earth's mass—and the Sun, although 332,000 times more massive than Earth, is more than 400 times more distant from Earth than is the moon. These same tidal interactions take place wherever planets have satellites orbiting nearby.

Orbiting closely around Jupiter are four natural satellites, each about the size of Earth's moon. These satellites—called Io, Europa, Ganymede, and Callisto—are known as the Galilean satellites, because they were first sighted and described by Galileo in 1610. The tidal forces on these small bodies orbiting close to Jupiter are much greater than the tidal forces on Earth or the moon, because of the proximity and size of Jupiter, the largest planet in the solar system—the eight-hundred-pound gorilla of the neighborhood. These tidal forces distort and heat the interiors of the Galilean satellites, enough to melt rock on the closest, and to melt ice on the next.

The flexing and heating of these satellites by Jupiter can be envisioned in a simple experiment. Take a metal coat hanger, the kind that is shaped like a triangle with two sharply angled corners. Grasp the hanger by the long sides leading to one of the corners, and move the pieces apart and back together rapidly, thus flexing the corner. With only a few rapid flexes, the corner will get very hot to the touch. The energy that was expended in flexing the corner was converted to heat where the metal was being distorted. The tidal forces exerted by Jupiter on its nearby small satellites have the same effect on the satellite interior—the tidal energy is converted to heat.

Io is the closest to Jupiter, orbiting at a distance where the tidal heating is so intense that it melts rock. Volcanoes erupt at many places over its surface almost continuously. Europa is the next Galilean satellite of Jupiter, not as close to the giant planet as Io, but still close enough to show the effects of tidal flexing and heating. Europa displays much evidence of crustal resurfacing, particularly in the almost complete absence of craters on its smooth, icy surface. The water is not supplied externally, as with a Zamboni on an ice rink; it comes from a zone of melting ice far below the surface, and reaches the surface through major fissures and fractures in the icy crust. The outermost of the Galilean satellites is Callisto, beyond the reach of intense heating by Jupiter. Its icy surface displays the dense impact cratering of the early days of the solar system, with no evidence of a subsequent resurfacing that would indicate crustal melting and the presence of water.

Europa is a target in the search for extraterrestrial life, because of the likely existence of water in its interior.[9] But could life develop deep within Europa, without benefit of the life-giving energy of the Sun? If Earth can serve as an example, the answer is certainly yes. At the bottom of Earth's oceans, along the seams of the tectonic plates, bizarre but vibrant biologic communities have evolved in total darkness, energized completely by hot springs emanating from the oceanic crust. And on land, within caves near the surface, evolution has produced organisms without eyes, endowing them with other sensory organs that let them navigate their dark environment. NASA has already conceptualized a mission to Europa that envisions a robotic penetrator that could reach the zone of liquid water—but that might not even be necessary. If the resurfacing of Europa has used interior water, it may be that evidence of deeper life is sitting at the surface, frozen in the newer ice that has given Europa its smooth surface.

Saturn, the ringed planet beyond Jupiter, also has a host of satellites that display icy surfaces and are susceptible to tidal flexing and heating. One of these satellites, Enceladus, has recently revealed extraordinary features that hint at the possibility of primitive life getting a start in its interior. The unmanned *Cassini* spacecraft, operating in the vicinity of Saturn since 2005, has photographed large cracks in the ice near Enceladus's south pole, and instruments aboard *Cassini* have detected evidence of active H_2O venting from the cracks, in geyser-like features that also yielded carbon dioxide, nitrogen, and methane.[10] Even closer flybys of Enceladus, one at only fifteen miles above the surface, detected traces of other hydrocarbons such as acetylene, ethane, propane, benzene, and even formaldehyde. Clearly, the interior of Enceladus is an environment where organic molecules can form, and therefore it, too,

9. R. Pappalardo, J. Head, and R. Greeley, "The Hidden Ocean of Europa," *Scientific American*, October 1999.
10. C. Porco, "The Restless World of Enceladus," *Scientific American*, December 2008, 52–63.

may be a home to microbial life. Our solar system is proving to be far more diverse than ever imagined.

WATER AND LIFE BENEATH EARTH'S ICE

One need not travel to the far reaches of the solar system to seek life deep beneath ice. There are opportunities right here on Earth. The Antarctic ice blanket is on average about a mile and a half thick—at its thickest it is more than two and a half miles, about twelve Empire State Buildings stacked atop one another. And even though the surface temperature averages about −50° to −60° Fahrenheit over the year, at the base of this ice pile the temperature is warm enough to melt the ice. The heat comes from deeper within Earth, and although it is only a trickle of heat compared to what the Sun supplies, over time it has been enough to melt the base of the ice. Where does that meltwater go? Effectively, nowhere—it fills in low-lying topography in the rock surface to form what are called subglacial lakes.

Two and a half miles beneath the Russian Vostok scientific station sits the largest subglacial lake in Antarctica—Lake Vostok, about the size of Lake Ontario, and on average about a thousand feet deep. The ice beneath Vostok Station has been drilled and cored to a depth just a hundred or so feet above the lake surface, and at that depth the ice is about 450,000 years old.[11] That means that if there is any life in the lake, it has been isolated from life elsewhere on Earth for nearly a half million years. Just as Australia, because of its long isolation as an island, has many animals unique to the territory—the kangaroo, the koala bear,

11. We speak much more about the climatic history revealed in the Vostok ice core in chapter 6, but for the moment it is enough to realize that Lake Vostok has been covered with ice for a long time.

the platypus—so might Lake Vostok show some microbial evolutionary products that reflect its long isolation.

This presents a very interesting experiment—drill through the last remaining ice to reach the lake, draw samples of the water to examine for life, and compare what is found to life forms in analogous settings elsewhere. But the challenge of this experiment is to ensure that no present-day life forms from the surface are introduced into the lake by the drilling process. Very careful thought has been given to how to achieve a clean entry, but full agreement is not yet at hand.

H_2O ON THE MOVE

H_2O on Earth is continually moving from one reservoir to another. Water evaporates from the oceans, and some of it falls as rain or snow on the continents. Some of the precipitation infiltrates the ground and moves slowly through the subsurface reservoir. Some flows overland in rivers, returning to the sea only a month or two after precipitation. Most snowfall lasts only a season on land before rejoining the ocean reservoir. But the snow that falls on the Antarctic ice cap, and is later compressed into a layer of ice, may remain on the great white continent for hundreds of thousands of years before it creeps back to the sea in one or another of the many glaciers that drain ice from Antarctica.

Over short intervals of time, there is an equilibrium between the relative sizes of the reservoirs, but over long periods of time, large transfers between the two biggest reservoirs, the oceans and ice, occur. When water leaves the oceans to become a temporary icy resident on the continents, the sea level falls, the continental shelves are exposed, and the continents experience an ice age. The spread of ice over the land surface overrides the grasses, plants, and trees, and eliminates the vegetable component of diet for a wide variety of omnivores and herbivores. When the climate ameliorates and ice sheets melt, the water returns to the oceans, the sea level rises, and the newly exposed land surface is reoccupied biologically. These temporary loans of ocean water to the continents in the form of ice have taken place some twenty times during the past three million years of Earth's history alone, and several other times in the more distant past.

The fact that ice ages come and go tells us that ice on Earth is always on the cusp of existence—a push one way, and ice grows and spreads; a push the other way, and ice retreats and disappears. There is also a temporal component to the cusp with present-day relevance—on one side of the cusp are the ice ages of the recent past; and on the other side, the ascendancy of Earth's human population as a major player in the global climate system, a force that is pushing ice toward disappearance. Today we are perilously poised on the peak of that cusp.

CHAPTER 3
WHEN ICE RULED THE WORLD

Three feet of ice does not result from one day of cold weather.

—CHINESE PROVERB

Some 120,000 years ago, in what would one day be known as Finland, caribou grazed in the waning days of an unusually cool summer. The previous winter had delivered heavy snowfall, and as the brisk winds and shorter days of fall set in, there were pockets of last winter's snow remaining in sheltered crevices and shadowed valleys. This residual snow gave a head start to the next annual white blanketing of the land, and reflected some of the Sun's rays back to space even earlier than usual, thus discouraging those occasional warm fall days before the onset of winter. And that next winter lasted a little longer as well, and springtime melting got a late start; at the end of the foreshortened summer there was even more residual snow to turn away the radiant energy of the autumnal Sun. It did not take many such downward-spiraling years to yield summertime snow cover over the entire region, forcing the caribou and the woolly mammoth to find new grazing farther south. The growth of an ice sheet had begun, one that would, over the next hundred thousand years, cover much of the land of Europe and North America with a blanket of ice two miles thick, and freeze the surface of high-latitude ocean water.

At its maximum extent the ice covered Canada, Greenland, Iceland, and Scandinavia completely. Most of the British Isles, Germany, Poland, and Russia, all the way to the Ural Mountains and beyond into the West Siberian Plain, were beneath the ice sheet. The ice extended into what is now the United States as far south as the modern Missouri and Ohio rivers and east over New England. High mountains beyond the margin of the ice sheets—the Rockies and Sierra Nevada in North America, the Alps and Pyrenees in Europe, and the high ranges of Asia—also developed glaciers.

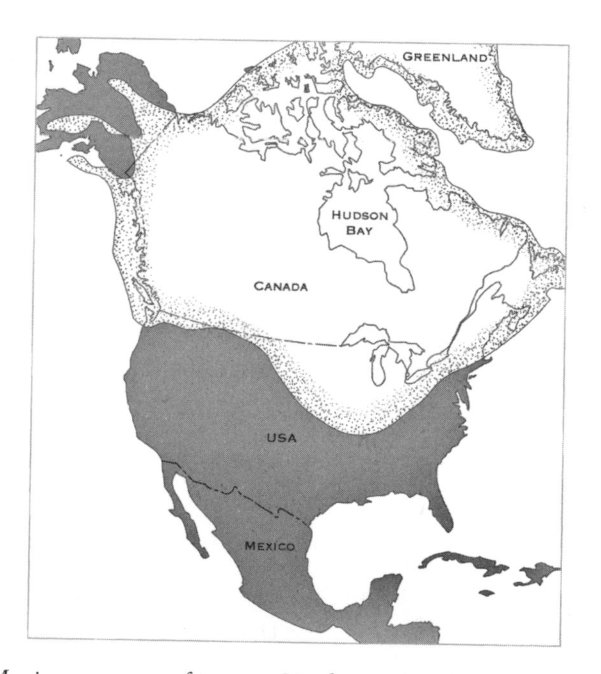

Maximum extent of ice over North America during the most recent ice age, from about 120,000 to 20,000 years ago

In the Southern Hemisphere, Antarctica was completely blanketed, and the high peaks of the Patagonian Andes, in South America, and on the South Island of New Zealand were strongly shaped by ice that flowed all the way to the sea. Even in Africa, astride the equator, Mounts Kilimanjaro and Kenya and the peaks of Uganda's Ruwenzori Range hosted extensive ice.

At sea, ice was widespread as well. More or less permanent sea ice covered the entire Arctic Ocean and reached all the way to Iceland, in the Atlantic. In the Pacific, ice extended south into the Bering Sea, between Alaska and the Russian Far East. The Southern Ocean surrounding Antarctica also had a year-round icy surface, reaching northward to the sixtieth parallel of south latitude. Seasonal sea ice in both hemispheres extended the range of ice even farther, but it is unlikely that the Drake Passage, between the Antarctic Peninsula and South America, ever was completely closed, because the strong winds and currents blowing through there did not let ice take hold.

Thick piles of ice, widely spread over Earth's surface just twenty thousand years ago—that is quite a concept. In fact, it was not until the late nineteenth century that scientists embraced the idea of widespread ice in the recent geological past. And geologists have since learned that this was only the most recent great ice excursion to spread over the Earth—one of a score or more that rhythmically advanced and retreated during the last three million years.

WHAT IS IT that geologists see that leads them to envision such a different world, a time in the history of Earth when there was much more ice on the planet than there is today? The evidence comes from both the land and sea, underfoot and underwater. On the land are multiple signatures in the landscape of an earlier presence of ice, and in the sea is evidence of the complementary signature—a reduction in water. The latter tells where the H_2O came from; the former tells where it went.

CLUES FROM THE LAND

When agriculture expanded from its beginnings in the Fertile Crescent of Mesopotamia into the plains of central and northern Europe, settlers found a terrain strewn with rocks and boulders, of all sizes, shapes, and colors. It was as if a giant spice bottle had been shaken overhead and had laid down a motley layer of pepper and paprika, sage and saffron, cloves and cinnamon over the ground. These silicate mega-spices were embedded in fertile silt and clay, and were so abundant and presented such formidable obstacles to plowing and planting that farmers were forced to clear boulder fields in order to gain enough area for crops. When Europeans came to the New World, they discovered that much of the northern United States and Canada was similarly strewn with a blanket of debris. Early farmers worked long days to move the boulders to the periphery of the fields or to stack them along the property line between neighbors, thus providing the setting for Robert Frost's poem "Mending Wall," with the memorable phrase "Good fences make good neighbors."

These boulders, heterogeneous in size, shape, and composition, are very different from the sediment in other geological layers. The mechanism for transporting sediment most familiar to early geologists is running water, which tends to segregate materials of different sizes—moving small-size particles along while leaving larger and heavier pebbles and boulders behind. Moreover, the tumbling of pieces of rock in a streambed tends to break off the sharp corners, eventually leaving the stones very well rounded. Similar rounding occurs on ocean beaches, where the incessant pounding of the surf produces well-rounded grains of sand. No, this blanket of mixed-up rock, of all sizes and shapes, was not laid down by water. It was simply dropped by whatever transported it so widely—and even "widely" is an understatement. This blanket of rock rubble can be found draped over some six million square miles on three continents of the Northern Hemisphere.

Ice was eventually recognized as the distributor of the debris. Unlike moving water, which sorts, rounds, and winnows the rock fragments it encounters, ice does none of these things. The key to understanding how ice picks up and delivers rocks to new locations was found by observing mountain glaciers, such as those seen in the European Alps and the Rocky Mountains of North America. But first, a few explanatory words about the mechanics of mountain glaciers: As snow accumulates in high areas year after year, the deeper layers get compressed into ice by the newer snow above. As the ice thickens, it slowly spills out of its catchment basin and begins to creep downhill, in a river of ice that flows a few tens of feet each year, truly at a "glacial pace." The ice descends to lower elevations, where it meets warmer air, and at some point the temperature reaches the melting point of ice. Beyond that point the glacier is steadily transformed into an ever-growing stream of meltwater.

As glacial ice descends from the heights, it erodes whatever bedrock it encounters, plucking and scraping rock from the walls and floor of the valley through which it flows. The debris is rafted along with the flowing ice

Large glacier spilling off the south polar plateau through the Transantarctic Mountains, in Antarctica

to lower elevations. The load is transported to the terminus of the glacier, that is, the place where ice melts as fast as it is delivered from above. At the terminus, the rock is deposited as an arcuate mound called a terminal moraine. The ice front, the snout of the glacier, may seem to be stationary, but in fact newer ice is always moving to the front, to meet the ultimate fate of melting. On its flow to the front, the glacial ice continues to carry rock debris for delivery to the moraine. The process is somewhat like placing suitcases on a descending escalator—they are transported to the bottom, where they are unceremoniously dumped, and simply pile up. As the glacier melts, the rocks are abandoned as a jumble of rough-around-the-edges newcomers from the heights, now relocated to a lower landscape. If the climate warms, the ice front will melt back farther up the valley, but the terminal moraine will remain where it was deposited, offering testimony to the greater reach of the glacier in earlier, colder times.

LIKE A HOT KNIFE THROUGH BUTTER

Ice is a powerful shaper of the landscape, and not only by conveying rock from one place to another. The great boulder fields strewn widely over large areas give testimony to the vast extent of the ice sheets, and today's mountain glaciers demonstrate the ability of ice to transport rock from myriad source areas. As impressive as those characteristics of ice are, the description of ice streams and ice sheets is incomplete without drawing attention to the titanic erosive power of moving ice. It is the power to excavate, bulldoze, break, crush, and pulverize rock as it moves over the terrain, the power to sculpt mountains and carve out valleys. There is almost nothing in the terrain that can withstand the prolonged passage of ice or forestall its reshaping of the landscape.

As ice succumbs to the pull of gravity and is drawn downward, it assaults the mountaintops where it has accumulated. The slow but persistent flow of ice in many directions away from the summit pulls rock from all sides, sharpening the top into a pointed and angular shape, recognized in the descriptive names humans have given them: the Matterhorn, the Beartooth, the Sawtooth. Once rock has been trapped beneath moving ice, it acts like coarse sandpaper on the bedrock below it, gouging and grooving the bedrock, leaving behind long striations that indicate the direction in which the ice moved. These scratches can be seen on glaciated terrains everywhere, from Glacier National Park in Montana, to the Upper Peninsula of Michigan, to Central Park in New York City, a clue that ice passed over these places.

But running water and even wind also modify the landscape, so geologists have had to observe carefully to identify characteristics unique to each type of landscape. The deep main valley of Yosemite National Park in California is U-shaped in cross-section, a characteristic of erosion by ice, in contrast with V-shaped valleys, which have been cut into mountainous terrain by streams of water, such as in the headwaters of the Ganges and Indus rivers high on the Tibetan Plateau.

Erosion by ice and water differ in other respects. Water is exceptionally agile, and can find and follow low pathways in virtually any topography. It can zigzag through the terrain, giving rise to valley systems that are anything but straight. Ice, on the other hand, is like a big, stiff, lumbering giant, making its way downhill with much less meandering. Rather than go around obstacles in the terrain, as water does, ice just plows through them, straightening, smoothing, and widening the terrain like a bulldozer blazing a new highway. The resulting valleys carved by ice have steep sides and smooth, round bottoms—the U shape mentioned earlier. They are deep, and offer a long and unobstructed view. The Finger Lakes of upstate New York, so named because of their long and narrow configurations and parallel development, are examples of valleys scoured and straightened by ice.

When valley glaciers erode deep U-shaped valleys all the way to the sea, the trough will become a fjord when seawater enters the valley after the ice has melted. Given the ample depth of fjords, oceangoing vessels can cruise

U-shaped valleys carved by mountain glaciers,
Torres del Paine National Park, Chile

tens and sometimes hundreds of miles "inland" along these valleys. These elegant topographic features are special remnants from the recent ice ages. They do not occur just anywhere; they form under only a certain set of conditions. A look at a world map, one that shows enough topographic detail, including fjords, reveals that they are well developed only along the west coasts of panhandle Alaska, Canada, Greenland, and Norway in the Northern Hemisphere, and along the west coasts of southern Chile and the south island of New Zealand. What do these places have in common?

First, they all are located where the prevailing winds generally pass over long stretches of ocean before reaching land, picking up substantial moisture along the way. As the wind encounters the higher and cooler elevations of the land, it gives up its cargo of water vapor as precipitation. Second, these "fjorded" coastlines are all at latitudes sufficiently far away from the equator (about 55° north or 45° south) so that during the recent ice ages, when Earth was colder by some ten to fifteen Fahrenheit degrees, the precipitation would fall as snow, and accumulate to form glacial ice. Third, at those latitudes or even farther toward the poles, it is cold enough even at sea level for glaciers to reach the sea instead of melting away while still flowing overland. Finally, because sea level is lower during an ice age, the glaciers flowed beyond today's high sea level coastline, and continued to carve deep valleys across the temporarily exposed continental shelves. When sea level later rose at the end of the ice age, as meltwater returned to the sea, the deep glacial valleys became inundated, giving us these long scenic waterways. Fjords are truly a gift of the ice ages, and a reminder of the tremendous erosive power of flowing ice.

ICE TO WATER

The piles of debris carried and ultimately dropped by the ice sheets created an irregular and rumpled surface. When the ice melted, the earth-moving ceased. The terrain was simply left as a work in progress, much like a road construction site on a weekend or holiday. But the glacial "workforce" left the site, never to the return, at least not for tens or hundreds of thousands of years. The low places filled with water. Many thousands of small lakes dot the debris-blanketed surface of Minnesota, Wisconsin, Michigan, and the Canadian provinces to the north. In Europe there is an equivalent "land of lakes" in Finland, parts of Sweden, and the far northwest of Russia. Flying over the lake-dotted landscape late in the day, one sees the low Sun reflecting off the surface of these lakes, little jewels glistening as far as the eye can see.

The effects of glaciation go well beyond the areas overridden by ice. If the geographic extent of ice cover on land is defined by the deposits of glacial debris and the lakes nestled within, the extended reach of glaciation can be seen in the water produced as the ice melted. The volume of water tied up in the continental ice sheets is immense—major glaciations withdraw enough water from the oceans to lower sea level by six hundred feet the world over. As the ice melted, streams of water transported and deposited sediment to form the many local sand and gravel quarries now exploited in construction, road building, and manufacturing. Quite in contrast to the unsorted character of the glacial deposits, the sediment in water-laid deposits is much more uniform in size, shape, and composition.

Meltwater also accumulated to form large lakes in areas peripheral to the ice. The North American Great Lakes and the Great Salt Lake of Utah are the diminished remnants of the last big melt-off, elements of today's landscape that give testimony to the earlier widespread ice. Great Salt Lake was once much larger, deeper, and fresher. The broad depression in the Great Basin of Utah received meltwater from the ice cover of the Rocky Mountains, to form what geologists call ancient Lake Bonneville. At its highest stand about seventeen thousand years ago, the surface of Lake Bonneville was about one thousand feet above the level of the present-day Great Salt Lake, and its water spread over much of western Utah, an area some twelve to thirteen times greater than today's Great Salt Lake.[1]

Why is Great Salt Lake saline, while the Great Lakes are fresh? The difference arises because Lake Bonneville occupied a closed depression from which there was no outlet—the only loss of water possible was through evaporation. Over most of the time since the highest level of Lake Bonneville, evaporation has exceeded precipitation and stream flow into the lake. The result has been a progressive decrease in lake level and a growing concentration of dissolved salt in the diminishing water volume. Large areas of the former lake bed of Lake Bonneville are now exposed, and reveal a "pavement" of salt that makes up today's Bonneville Salt Flats, where racing cars set world speed records in excess of six hundred miles per hour. On the sides of the mountain ranges that rise out of the Great Basin one can easily see older shorelines of Lake Bonneville, like giant bathtub rings, recording pauses in the fall of the lake level.

The five Great Lakes of North America—Lakes Superior, Michigan,

1. Much interesting material about Lake Bonneville and Great Salt Lake can be found at http://geology.utah.gov/online/PI-39, a site of the Utah Geological Survey.

Huron, Erie, and Ontario—border eight states and Canada. Together they contain 85 percent of the surface water of North America, and 20 percent of the world's freshwater. They are all connected by rivers—the St. Marys River carries water from Lake Superior to Lakes Huron and Michigan; the Detroit River links Lake Huron to Lake Erie; and the Niagara River connects Lake Erie to Lake Ontario, with one giant leap over Niagara Falls. Lake Ontario empties into the St. Lawrence River, the final long waterway that delivers the Great Lakes water to the Atlantic Ocean.

The depressions that these lakes occupy were shaped by the North American ice sheet, which scraped away at weak bedrock to create low regions that would later host the lakes. The Lake Michigan and Lake Huron basins, which nearly surround the lower peninsula of Michigan, sit in relatively weak rocks, Paleozoic shale that was no match for a mile-thick ice sheet grinding its way south. By contrast, the Niagara dolomite, a very tough rock formation composed of magnesium carbonate, stood up better to the erosive power of the ice. The Niagara formation forms a sweeping arc around the Great Lakes—it is the backbone of Wisconsin's Door Peninsula, behind which lies Green Bay, and of Manitoulin Island and the Bruce Peninsula of Ontario, which separate Georgian Bay from Lake Huron. As the name Niagara suggests, this rock layer also forms the durable platform on which the Niagara River flows out of Lake Erie—its lip creates the ledge where the river plunges 190 feet at Niagara Falls, on its way to Lake Ontario and the sea.

Although the lake basins resulted from glacial erosion, the water now filling them is not glacial meltwater; the water in the lakes has been replaced many times since the end of the last ice age. Rain and snow falling each year in the upper Great Lakes catchment basin replenishes the water that flows over Niagara Falls, through Lake Ontario, and outward to join the Atlantic Ocean. The annual loss and replenishment is about 1 percent of the volume of water in the lakes, so it takes about a hundred years to totally exchange the waters in the Great Lakes—pollution introduced into the lakes takes a century of flushing to purge.

There are similarities and differences between the Great Lakes and Lake Bonneville of Utah. As with Lake Bonneville, which had bigger volumes and higher lake levels in the past, so have the Great Lakes had higher stands. But unlike Lake Bonneville, the Great Lakes basins were covered by the North American ice sheet at the time of its maximum extent, some twenty thousand years ago. As the ice front retreated northward from the region, for a time the ice actually formed a barrier that prevented meltwater from exiting through the St. Lawrence River. The meltwater, blocked by the ice from flowing northeast, instead found its way into the Mississippi River and the Gulf of Mexico. Effectively the ice was a dam along the north margin of the Great Lakes, and for a while it led to higher lake levels than exist today.

Satellite photos reveal several former shorelines along the margins of Lakes Michigan, Huron, and Erie. Each of these lakes had spread over a greater area than they occupy today—and the relict lake beds, flat and blanketed with fine sediment that settled out from the ancient lakes, are put to good use. These large, level, and featureless plains make for easy use in agriculture, and are attractive sites for airports. Detroit's Metropolitan Airport is located on the vast flat exposed lake bed of an earlier and bigger Lake Erie, twenty miles away from the western shore of today's Lake Erie.

Another large meltwater lake once covered much of the Canadian province of Manitoba, but extended also into Ontario, Saskatchewan, North Dakota, and Minnesota. This lake, named Lake Agassiz after the nineteenth-century Swiss geologist Louis Agassiz, was big—some seven hundred miles north to south and two hundred miles across. At its maximum extent some thirteen thousand years ago, it was bigger than either California or Montana, almost two thirds the size of Texas. Its water covered an area 80 percent greater than all of the modern Great Lakes combined. This was the great lake of its time.

But Lake Agassiz returned most of its glacial meltwater to the sea. Abandoned shorelines on hillsides and a giant exposed lake bed across the plains of southern Canada and adjacent north-central United States provide evidence of the lake's former extent. Today its much-diminished remnant is Lake Winnipeg—between Lake Erie and Lake Ontario in size—big by today's standards, but a shadow of its former self.

As the ice sheet melted, big rivers developed to drain the immense volume of meltwater and return it to the sea. The ultimate margin of North America's last great ice sheet can be identified by the major rivers established on the periphery. Today we know them as the Missouri and Ohio rivers, along the southern boundary of the ice sheet, and the Mackenzie River to the west. The Mackenzie flowed northward to the Arctic Ocean, but the Missouri and Ohio joined the Mississippi to drain much of the early meltwater to the Gulf of Mexico. As the ice sheet melted back, other outlets to the sea opened. When the St. Lawrence River began to drain the Great Lakes Basin, the meltwater no longer coursed south to the Gulf of Mexico—it was delivered to the North Atlantic Ocean, with profound, albeit temporary, consequences to the Atlantic circulation and climate.

LOUISIANA OF THE NORTH

In the early nineteenth century the Ohio and Missouri rivers were the pathway to the interior of the continent for the explorers Meriwether Lewis and William Clark. The Lewis and Clark Expedition had been authorized by President Thomas Jefferson shortly after he concluded the big territorial acquisition known as the Louisiana Purchase in 1803. Jefferson had purchased from Napoleonic France a huge swath of land in the interior of North America west of the Mississippi River, land that today includes all or part of fifteen states, representing almost a quarter of the area of the United States. The Lewis and Clark journey of 1804–6 had a simple purpose: to see what it was that we had just acquired from France. Wrote Thomas Jefferson:

> The object of your mission is to explore the Missouri river, and such principal stream of it as by its course and communication with the

> waters of the Pacific Ocean whether the Columbia, Oregon, Colorado or any other river may offer the most direct and practicable water communication across this continent for the purposes of commerce.

The scholarly Jefferson was interested in much more than simply commerce, and he instructed Lewis and Clark to make friendly contact with native peoples encountered along the way, and to observe the flora, fauna, and mineral resources of the different regions. The expedition followed the Missouri River to its headwaters along the present-day Montana-Idaho border, and then crossed the Continental Divide, the great watershed that separates westward drainage to the Pacific from waters headed eastward. There they joined the Snake River and followed it to the Columbia River, their pathway to the Pacific Ocean.

The Columbia River basin owes much of its landscape to the outflow of glacial meltwater. The Columbia crosses both Washington and Oregon, flowing over and through a rock formation known as the Columbia River Basalt. These are volcanic rocks—lava flows that spilled over much of the Pacific Northwest during the mid-Miocene period, around fifteen million years ago. On top of the basalt sit a few hundred feet of wind-blown dust. This was the terrain that at the end of the last ice age was exposed to one of the most catastrophic floods in human history.

The scenario for this massive flood, which came to be known as the Spokane, or Missoula, Flood, began with the formation of a temporary ice dam in the narrows of the Clark Fork River, in western Montana, around fifteen thousand years ago. This dam caused a large lake, Lake Missoula, to form behind it, in the same way that the ice front in the mid-continent led to temporarily higher levels of the Great Lakes. When the ice dam failed suddenly, the impounded lake water burst into the drainage ways that fed the Columbia River, and poured downriver with a velocity nearing fifty miles per hour. Some estimates of the volume of water that cascaded over the terrain suggest that it exceeded the total flow of all the rivers of the world, at least for a few days.

This torrent shaped the landscape in extraordinary ways. It cut deep canyons into the basalt caprock, leaving occasional large escarpments that drop from one basalt flow to another. The lips of these escarpments display a scalloped shape, much like the Horseshoe Falls at Niagara. Indeed, giant waterfalls did cascade over these cliffs, carrying immense volumes of water that scoured deep plunge pools at the cliffs' base. Large tabletop "islands" capped with basalt, similar to the mesas in the desert landscape of Utah and Arizona, stand isolated by channels scoured around them. This churning, turbulent sheet of water eroded huge boulders of basalt, much bigger than a house, tumbling them miles downstream and eventually dropping them in a plain marked by gigantic ripple marks, so large that the rhythmic rise and fall of the topography can be fully appreciated only from the air.

The surge of water continued toward the Pacific coast. It cut a canyon across the narrow continental shelf, at that time still exposed because of the lower sea level of the ice age. When the flood entered the sea, it dropped its sedimentary load on the ocean floor. Some of the debris had come all the way from Montana. And if one such flood was not sufficiently cataclysmic, geologists suggest that this scenario was repeated many times over the next two thousand years. Eventually, the ice retreated far enough north so that ice dams no longer formed in Montana, and outbursts no longer washed over the Columbia Plateau.

This extraordinary terrain arrayed across Washington is called the Channeled Scablands. The scale of the features in the landscape and of the processes that formed them—like continent-wide ice sheets themselves—is beyond anything in the modern human experience. These landforms stand there today—the channels carry little water, the basalt boulders tumble no more, and the falls are dry. They offer mute testimony to a time when glacial meltwater roared across this lava plateau in a massive flash flood.

Archeological evidence now documents a human presence in Oregon some fourteen thousand years ago, perhaps the earliest humans ever in western North America. Imagine the reaction of these early beings to the periodic walls of water that churned through the valleys and spread catastrophically across the landscape. The estimates of the recurrence interval of the floods are around fifty to sixty years, well within the lifetimes of early residents of the region. These floods would be the stuff of legend and oral history, passed on through many generations, similar to the historical memories of major tsunamis that have affected coastal populations in earthquake-prone regions.

When J. Harlan Bretz, a geologist at the University of Chicago, first offered the giant flood hypothesis in the early 1920s, he became an object of derision. What he had proposed was an example of catastrophism in the geologic record, a concept that gave special emphasis to the role of infrequent and improbable events in the shaping of Earth's surface. The conventional geological wisdom of the time favored the concept of uniformitarianism, whereby the landscape could be interpreted in terms of observed and reasonably well understood processes acting slowly over long periods of time. But Bretz persisted in offering more and more field evidence to support the flood concept, and eventually his perspective was accepted.

THE THICK PILE of ice sitting on the surface of North America and Eurasia had another remarkable effect—it formed such a massive load that the crust of Earth actually sagged beneath it. In central Canada the depression was sufficiently deep that when the ice melted and sea level rose, ocean water filled the depression, producing the large marine embayment now called Hudson Bay. Today, with the load of the ice gone, the crust is slowly rebounding upward, and spilling the seawater in Hudson Bay back into the deeper ocean basins. Just as Lake Bonneville became smaller because of evaporation, Hudson Bay is also shrinking as more and more of the continental crust reemerges above sea level.

In Europe, the center of the ice pile was in the Gulf of Bothnia, between Sweden and Finland. As in North America, Earth's crust was depressed beneath the load of ice, and following the melting of the ice, the rock surface is slowly emerging from beneath the sea. These long-term geological processes are slowly adding territory to the maps of Canada and Fennoscandia (the geological name for the Scandinavian peninsula), a peaceful restoration of land following the invasion and occupation by the glacial ice. But whether the restoration will continue is not so clear. Post-glacial rebound will continue to uplift the still-depressed crust, but the uplift is now meeting competition as anthropogenic climate change diminishes polar ice and raises sea level. If sea level rises faster than the crust does, the sea may once again conquer the land.

ICE AGES LEAVE THEIR MARK IN THE SEA

In the ocean, the signature of an ice age can be found in the chemistry of seashells that form layers on the ocean floor. As water evaporates from the ocean and falls as snow on the growing continental ice sheets, the chemicals dissolved in the remaining seawater become more concentrated. Since small marine creatures use the ocean's chemicals to grow their shells, the composition of their shells during an ice age reflects the more concentrated chemistry of the oceans in which they grew. When these creatures died and fell to the ocean floor, the shells accumulated as a layer of history representing a time when higher chemical concentrations indicated less water in the oceans. These stratified marine cemeteries reveal that no fewer than twenty ice ages have rhythmically paraded over the continents in the past three million years, each time borrowing water from the oceans, which forced the remaining seawater to carry a heavier chemical burden.

The particular signature of water withdrawal is in the relative abundance of one of the uncommon isotopes of oxygen: ^{18}O. This isotope of oxygen has two extra neutrons in its nucleus, making it about 12 percent heavier than the very common isotope ^{16}O. That extra weight makes it harder to vaporize (evaporate) water containing ^{18}O as compared to ^{16}O, with the result that as water leaves the ocean on loan to continental ice,

^{18}O becomes more concentrated in the remaining seawater, leading to an increase in the ratio of $^{18}O/^{16}O$. Consequently, the marine creatures that use the chemicals in the ocean for shell construction will record this higher concentration of ^{18}O in their shells, which in turn is the signal of less ocean water and, by implication, that ice volumes are higher. Very clever indeed.

Under the shallow waters that now cover the continental shelves, other topographic features also indirectly point to lower sea levels. Along the eastern coast of the United States, off the present-day mouths of the Hudson River and of the Susquehanna River emptying through Chesapeake Bay are deep valleys cutting a few hundred miles across the continental shelf into the Atlantic Ocean Basin. When sea level was lower during the last ice age, the continental shelves were exposed, and these rivers had farther to go to reach the sea. As the ice began to melt, the rivers carried much more water and had greater erosive power than they do today. This enhanced river discharge and strong erosion across the exposed continental shelf cut deep valleys, which can be imaged with marine geophysical methods, the same tools that are used to explore for oil in the rocks of the continental shelf. Today these valleys, now completely submerged beneath the ocean surface, are called the Hudson Canyon and the Baltimore Canyon.

At the edge of the continental shelf, the submarine topography drops steeply into the true ocean basin. Most of the water in the oceans of course resides in these low-lying areas. The fact that today there is also ocean water atop the continental shelves is a statement that there is more water on Earth than the ocean basins can accommodate, so some of it laps onto the margins of the higher-standing continents. In geological terms, oceans are defined not by where marine waters are found, but rather by the deep basins that surround the elevated continents. Indeed, if all the ice now present on Earth were to melt, sea level would rise onto the continents another 250 feet above today's level.

At the maximum of the last ice age, some twenty thousand years ago, sea level was lower than today by some six hundred feet, and the shoreline of eastern North America was at the edge of the continental shelf, in places several hundred miles away from today's coastline. Icebergs discharged into the Atlantic would float along that margin, and sometimes their deep keels would drag along the bottom, carving gouges into the rock and sediment. These long scour marks can be seen on the ocean floor today off the coast of South Carolina.[2]

WHEN MELTWATER REACHES THE SEA

In financial markets, transfers of capital between one market segment and another can be accommodated without upsetting the market unduly, provided the transfers are in small enough parcels spread out over reasonable periods of time. An "orderly" market follows such a pattern. But if huge blocks of stock in one sector are dumped onto the market all at once, turmoil can overwhelm the marketplace.

This financial analogy is apt for thinking about transfers between the ice and ocean reservoirs in Earth's hydrological system. As glacial ice melts in response to a slowly changing climate, meltwater forms streams that merge into rivers, and the rivers eventually reach the sea. And the freshwater of the rivers is gradually mixed into the saltwater of the oceans via the action of wind and ocean currents. The great hydrological accounting book will show the balance in the ice account slowly going down and the balance in the ocean account creeping up. In essence, there is an "orderly" shift of hydrological capital from one reservoir to another. On occasion, however, there can be abrupt shifts in the movement of hydrological capital that cause chaos in the exchange. One such moment of turmoil occurred during the melting of the last North American continental ice sheet.

2. J. C. Hill et al., "Iceberg Scours Along the Southern U.S. Atlantic Margin," *Geology* 36, no. 6 (2008): 447–50.

Just prior to the abrupt shift, the meltwater along the southern margin of the North American ice sheet was drained by the Missouri and Ohio rivers and delivered to the Mississippi River for final passage to the Gulf of Mexico. There it mixed with warmer and saltier water, eventually making its way into the open Atlantic Ocean, where it was stirred into the general circulation pattern of the Atlantic current system. But by about 12,800 years ago, the front of the North American ice sheet had melted back to a point where the meltwater found a new and shorter pathway to the sea—the topographic lowlands that would suddenly become the St. Lawrence River Valley. Large volumes of cold meltwater no longer went the Mississippi way, but instead coursed northeastward, past the Gaspé Peninsula and into the Atlantic Ocean. There it interjected itself into the slow northward drift of the Gulf Stream, inserting a cold, fresh barrier into this warm surface current, and interrupting the delivery of heat to the far North Atlantic. A big chill fell over the region, and people living farther to the north must have wondered what happened for the heat to be turned off so abruptly.

This cold period, which chilled Iceland, Greenland, and Western Europe for more than a millennium, has been called the Younger Dryas, because in the regions affected by the sudden chill, a cold-climate flower, the Dryas, was reestablished. It had grown previously in the regions a few thousand years earlier, when these latitudes were emerging from the last glacial maximum. The end of this thousand-year cold wave came when the meltwater was reduced to a volume inadequate to maintain a thermal barrier to the Gulf Stream, which then resumed its delivery of heat to the far north, accompanied by a substantial warming of Iceland and Western Europe.[3]

3. We will revisit the topic of climatic instabilities in chapter 8, which includes a discussion of tipping points in the global climate system. In particular, the question of destabilizing the Gulf Stream will be reexamined in the context of contemporary climate warming that is leading to the rapid loss of summer sea ice in the Arctic Ocean.

WHAT CAUSES AN ICE AGE?

Why do ice ages come and go? What factors lead to a periodic accumulation of thick ice on the continents? These questions are more difficult to answer than simply assembling the evidence that shows that ice ages have occurred. Because continental glaciations are relatively rare occurrences in the geological record, we know that the conditions producing ice ages do not occur very often. The previous occurrence of widespread glaciation (prior to the multiple ice advances of the past three million years) was in Gondwanaland during the late Paleozoic, about 275 to 300 million years ago.

Some generalities, however, can be garnered from the geologic record. One obvious factor that plays an important role in glaciations is the location of landmasses on the globe. Land that is situated at high latitudes, where it is colder, is a more probable setting for snow and ice accumulation. Today Antarctica lies entirely within the Antarctic Circle, and much of Greenland lies north of the Arctic Circle. Conversely, where there is only open ocean at high latitudes, as with the present-day Arctic Ocean that surrounds the North Pole, there will be no thick accumulations of ice on the sea surface. Sea ice, of course, does form over the Arctic Ocean, but it is a thin and fleeting cover, with marked seasonal variations in extent, and a lifetime measured in years or perhaps decades. By contrast, accumulated land ice can survive hundreds of thousands of years. The oldest ice in Antarctica is about eight hundred thousand years old,[4] and in Greenland, a little more than one hundred thousand years old.[5] Other factors, such as changes in the patterns of atmospheric and oceanic circulation, play a role in making

---

4. J. Jouzel et al., "Orbital and Millennial Antarctic Climate Variability over the Last 800,000 Years," *Science* 317 (2007): 793–96.
5. R. B. Alley, *The Two-Mile Time Machine: Ice Cores, Abrupt Climate Change, and Our Future* (Princeton, N.J.: Princeton University Press, 2000).

Earth susceptible to ice accumulation in the high latitudes. One such event (discussed in chapter 1) was the climatologic isolation of Antarctica brought about by the opening of the Drake Passage.

During the last three million years, conditions developed for "a perfect storm" that has led to multiple oscillations of ice over the planetary surface, with the most recent ice sheets retreating only ten to twenty thousand years ago. What happened three million years ago that set the stage for recurring accumulations and dissipations of ice? One important change was the tectonic uplifting of the Isthmus of Panama, to link North and South America. But more important than the connecting of two continents was the disconnecting of two oceans, the Atlantic and Pacific. Prior to the formation of Panama, the Atlantic and Pacific exchanged water via east–west currents that flowed through the gap between North and South America. But when Panama blocked that exchange, the Atlantic surface currents became predominantly south to north, and carried warm water into the Arctic, water that altered the atmospheric circulation and precipitation patterns in the far north. More snow began to fall, and not all of it melted in the subsequent summers. The modern cycle of ice ages had begun.

A very elementary question about the causes of ice ages focuses simply on the issue of accumulation: What conditions lead to winter snowfall greater than summer melt-off? When year after year more snow falls than melts, there is a growing accumulation—that is the scenario laid out at the opening of this chapter. It is analogous to your savings account—when year after year you spend less than you earn, there is an accumulation in the piggy bank. The conditions that favor an accumulation of snow are related to changes in the seasons on Earth, particularly at high latitudes, where it is already cold enough to make summertime melting a short-lived phenomenon. The crucial arena is near the Arctic Circle, in the Northern Hemisphere, which passes through Alaska just north of Fairbanks, across northern Canada and the southern tip of Greenland, near Iceland, through Scandinavia, across northern Russia, all the way to the Bering Strait. Anything that can slightly alter the competition between winter and summer, that leads to shorter, colder summers at that latitude, may trigger the onset of an ice age.

So what causes the seasons on Earth? Why do we have winter, spring, summer, autumn, and then another winter? There are two principal and unequal effects that lead to seasonality. The greater effect is called the tilt season, and the lesser, the distance season. The tilt season is associated with Earth's rotational axis, that imaginary line running through the planet from pole to pole, around which Earth spins daily (and 365 times plus a fraction in its yearly journey around the Sun). The rotational axis is not, however, perfectly upright relative to the plane of Earth's orbit about the Sun—it is tilted a little more than twenty-three degrees away from upright. That means that in the course of the year, first one hemisphere and then the other will get slightly more sunshine, because it is tilted toward the Sun. In the annual journey around the Sun, when a hemisphere is tilted toward the Sun, it experiences summer, and when it is tilted away from the Sun, it experiences winter.

The second and lesser cause of seasons relates to the fact that Earth's orbit about the Sun is not a perfect circle, but rather an ellipse, a slightly squashed circle with a long dimension and a short one. And the Sun does not sit exactly in the middle, but rather a bit off-center, closer to one end of the long dimension than to the other. This means that as Earth follows the elliptical path around the Sun, its distance from the Sun is always changing. As Earth comes closer to the Sun, it gets a little extra sunshine, and as it moves away from the Sun, it gets a little less. This variation in solar heating during every trip around the Sun also contributes to warmer and colder periods during the year—giving rise to the distance seasons.

The tilt season and the distance season combine to produce the actual seasonal variation in sunshine, but the sum is different in each hemisphere. When the Northern Hemisphere is tilted directly toward the Sun on June 21, Earth is at the most distant part of its orbit. So, while tilting gives the North some extra sunshine, the distance effect gives it a little less, thus diminishing the heating from the tilt. Six months later, on December 21, when the Southern Hemisphere is tilted directly toward the Sun, Earth is making its closest approach to the Sun, so the tilt and distance reinforce each other. This produces asymmetry in seasonality between the hemispheres—the Southern Hemisphere has stronger seasonal variation than does the Northern Hemisphere.

That is the basic picture of how a typical year's sunshine gets spread over the hemispheres to produce seasons, but it only shows us how seasons are established—it does not tell us what changes to this picture of seasonality would allow accumulations of snow at high latitude year after year, and initiate an ice age. That story is more complicated.

There are very small periodic variations in the amount of seasonal sunshine that the hemispheres receive through the tilt and distance effects, because the tilt of Earth's rotational axis and the elliptical shape of Earth's orbit about the Sun are themselves changing, albeit very slowly. These slight changes are imposed by the gravity fields of the other planets in the solar system, but the principal effects come from Jupiter, the largest and most massive planet in the solar system, some 318 times more massive than Earth.

There are three perturbations to Earth's seasonal story that result from these planetary gravitational tugs, one that affects the distance season, a second that affects the tilt season, and a third that determines the geography of where the interactions of the other two add to or subtract from each other. These slowly changing effects are called the Milankovitch cycles, in honor of Milutin Milankovitch, a Serbian geophysicist who early in the twentieth century pointed out their importance for climate change. In the longest of the Milankovitch cycles, the shape of the ellipse becomes slightly more elongated, then slightly more circular, and then back again, while at the same time the ellipse is slowly rotating around the Sun to trace out an orbit that over a long period looks like petals of a flower. A full oscillation in the elongation takes about 100,000 years, and changes the distance contribution to the seasons.

In a second cycle the tilt of Earth's rotation axis, currently about 23.4 degrees, oscillates between 22.1 and 24.5 degrees every 41,000 years. At greater tilts, the seasons become more extreme, and at lesser tilts, more uniform. A third cycle arises from the precession of the rotational axis—a familiar effect seen in the slow wobble of a spinning top's axis of rotation. This changes the orientation of the rotation axis, and determines the hemisphere in which the distance and tilt season reinforce each other, as in the Southern Hemisphere today. The precession of Earth's rotational axis makes a complete cycle with respect to the seasons in about 23,000 years—so, in half that time, the reinforcement will occur in the Northern Hemisphere, before shifting back south of the equator to complete the cycle. The maximum amplification of the seasonal contrast takes place when the orbit is most elongate and the tilt is at its maximum. The precession determines which hemisphere will receive the maximum reinforcement of the distance and tilt effects.

The effects at these three periods—100,000; 41,000; and 23,000—combine at any given time to produce a composite increment or decrement to the sunshine received at a given place on Earth. The composite is like listening to sound from an electronic synthesizer, which uses only three tones with different volume settings. The combination is usually some gentle cacophony, but from time to time there is some harmony between two of the tones, and on occasion with even one tone dominating, coming through loud and clear. The right combination of these Milankovitch factors sets the stage for snow accumulation at high latitudes and the beginning of an ice age.

Two lines of evidence suggest that for the last several hundred thousand years it has been the 100,000-year oscillation in the distance effect that is dominating the variations in sunshine and seasonality at high latitudes on Earth. One line is found in the sea, in the layers of marine fossils that indirectly provide the history of ocean water volume by way of the $^{18}O/^{16}O$ ratio. These sedimentary layers show four recent low-water stands separated by an average of just over 100,000 years. The second line of evidence comes from deep ice cores from the Russian Vostok borehole through the East Antarctic ice sheet. These cores show the same 100,000-year periodicity in the polar air temperature, with the lowest temperatures occurring at the same time that the oceans show their lowest sea level. It can be no coincidence that the coldest temperatures occur at the same times that the oceans display minimum water volume—the ocean water has been transferred to the land and frozen into an ice blanket, which is sitting on the cold continents. Earlier glacial cycles revealed in the Antarctic ice show the 41,000-year periodicity. This indicates that the oscillations in the tilt were then the dominant factor in the three-note tone poem composed by the bobbing and weaving of Earth in its long-term relationship with the Sun and other planets.

While it is clear that the Milankovitch climate cycles are the pacemakers of the ice ages, other factors come into play as ice accumulates. The albedo, or reflectivity, of Earth begins to increase, and vegetation is overridden by the spreading ice, thus altering both the polar radiation budget and carbon cycle. These changes amplify the polar chill initiated by the Milankovitch influences on seasonality.

HUMANS ON THE MOVE

The closure of the Isthmus of Panama three million years ago brought changes to more places than just the poles. The adjustments in the circulation of the oceans and the linked effects on the atmosphere led to redistribution of global precipitation. Africa, the cradle of human development and evolution, was no exception. An aridity set in that led to a cooling and drying of the continent, particularly north of the equator. Forests gave way to grasslands, and the Australopithecines, the progenitors of the modern human genus *Homo*, struggled to adapt. Evolutionary pressure favored new tools and skills. The emergence of stone tools marked the beginning of the so-called Paleolithic period of human evolution—and the makers and users of these tools were early representatives of the genus *Homo*. These early humans spread widely through Africa between 2.5 and 1.5 million years ago, and later into Europe and southwest Asia. There they learned—or, to their detriment, did not learn—to cope with a slowly oscillating climate, and with ice sheets that periodically moved across northern Europe and Asia.

The archeological site at Atapuerca, in northern Spain near the city of Burgos, is known as the home of the first Europeans,[6] with evidence of occupancy dating back to 1.2 million years before the present. Dr. Josep Pares is a member of the scientific team at the National Research Center on Human Evolution in Burgos (and my colleague at the University of Michigan). These anthropologists and geologists have for more than a decade been engaged in reconstructing the human history at this site. Pares's role has been to create a time frame for the human presence at Atapuerca, using a wide array of geological and geophysical techniques to determine the age of the occupancy. In 2007 he took me to the dig and guided me into a cool dark cave and through layer after layer of sediment containing the bones and fossils and tools that tell of this early distant outpost of humanity, on its journey out of Africa and to the far corners of the world.

For the most part, this journey was on foot. It is not easy to walk around the world, but as the ancient proverb says, "A journey of a thousand miles begins with a single step." And as various early representatives of our genus *Homo* left Africa, little by little they expanded into new regions of Europe and Asia, always in search of abundant food. In the last half million years the Heidelberg species of *Homo*, and later the Neanderthals, became established in Europe, while at the same time, back in Africa, another human species, *Homo sapiens,* was in ascendancy. Midway through the last ice age, some seventy thousand years ago, *Homo sapiens* felt the winds of climate change in Africa—a drying out that pushed

6. J. Eudald Carbonell, J. Pares, et al., "The First Hominin of Europe," *Nature* 452 (March 27, 2008): 465–69.

food gathering to an untenable situation and thinned human populations almost to extinction.[7] Thus began the final great human migration to all the habitable continents of the globe. *Homo sapiens*—our immediate genetic ancestors—were on the move to distant places. Their pathways around the world were opened, ironically, by the glaciation of the last ice age, a large-scale conversion of ocean water into continental ice.

What happens between water and ice that opens pathways for migration? Water and ice are two faces of the same material: H_2O. Earth's total endowment of H_2O is more or less stable, but during an ice age the proportions of the H_2O in ice and in water change. As noted in the earlier discussion of the isotopic chemistry of the oceans, large expanses of ice sitting on continents during an ice age represent large withdrawals of water from the oceans. One expression of the withdrawal is the lowering of sea level. This adjustment in the water budget of Earth leads to exposure of the continental shelves, and provides a dry-land human migration path that in more temperate times would be covered by shallow seas. The broad, shallow platform between Southeast Asia and the many islands of Indonesia, Papua New Guinea, and Australia provided a partial walkway of migration that led to human occupancy of Australia by fifty thousand years ago.

Another migration pathway of early *Homo sapiens* led eastward across Asia to the Kamchatka Peninsula and the Chukotka region of the Russian Far East, some thirty thousand years ago. The maximum of the most recent ice age was still ten thousand years in the future, and sea level would drop still farther. The peopling of the Americas was a direct consequence of exposing the seafloor across the Bering Strait and along the Aleutian Islands arc. The Bering route, some thousand miles across at its widest, brought people to the Arctic regions of Alaska, Canada, and into the mid-continent of North America by fourteen thousand years ago.

The flow of people along the Aleutian route went to southern Alaska

---

7. D. M. Behar et al., "The Dawn of Human Matrilineal Diversity," *American Journal of Human Genetics* 82 (2008): 1–11.

and southward along the western coast of North America. Evidence of humans in coastal North and South America[8,9] also dates from around fourteen thousand years ago. It's a long walk from Alaska to Patagonia, around nine thousand miles. But over a thousand years, it really amounts to a slow diffusion into new territory, some ten miles or so each year. There would be ample time for these folks to pick the berries and smell the roses, and marvel at the new world they were occupying for the first time. With the spread of humans into the Americas, the dispersal of *Homo sapiens* into all the habitable continents had reached completion.[10] And just in time, for the great melt-off of the glacial ice sheets had begun, and the land bridges from Asia created by lowered sea level were disappearing fast. The newcomers to the Americas were on their own.

The last ice age was nearing an end, and the ice front continued to retreat northward. Yes, there were interruptions to the fast warming of the climate: the sudden drainage of the large transient meltwater lakes—the Younger Dryas event that cooled the Northern Hemisphere for a millennium between 11,500 and 12,500 years ago, and another later, shorter drop in temperature 8,200 years ago, when Lake Agassiz emptied much of its water quickly into Hudson Bay and on into the Atlantic. But these were the last gasps of the ice age—temperatures around the globe reached levels similar to those of today, and remained more or less at that thermal plateau for the next 8,000 years. It was a remarkable span of climatic stability that enabled humans to flourish and multiply.[11] Sedentary agriculture largely replaced a nomadic lifestyle, and communities developed that recognized the advantages of occupational specialization.

---

8. M. T. P. Gilbert et al., "DNA from Pre-Clovis Human Coprolites in Oregon, North America," *Science* 320 (2008): 786–89.
9. T. D. Dillehay et al., "Monte Verde: Seaweed, Food, Medicine, and the Peopling of South America," *Science* 320 (2008): 784–86.
10. T. Goebel, M. R. Waters, and D. H. O'Rourke, "The Late Pleistocene Dispersal of Modern Humans in the Americas," *Science* 319 (2008): 1497–502.
11. Brian Fagan provides a very readable account of this period in *The Long Summer: How Climate Changed Civilization* (New York: Basic Books, 2004).

THE RECURRING ICE ages of the last three million years dramatically shaped the landscape, and stressed the small human population who, out of necessity, learned to cope with climate change and its consequences. Ice sheets overran vegetation that the early humans and their animal cohabitants both relied on. Water supplies were continually shifting. Encroaching ice forced humans to be on the go, and lower sea levels opened up avenues of human migration. Ice ruled the world, and humans simply reacted to its advances and retreats. By the end of the last ice age, the ingenuity of humans had been honed by the stress of multiple glaciations, and the human species, equipped with enhanced technical skills, was poised for a rapid growth in population.

The subsequent ascendancy of *Homo sapiens*, both in numbers and in capabilities, was beginning to leave a mark on the planet. The large land mammals, such as the mammoths and mastodons, became extinct, in part due to the pressures of human hunting—an early demonstration of what would come later, when the bison of North America were hunted nearly to extinction. And as human communities developed agriculture, the changing uses of land and water slowly altered the web of life in their environment. But little did anyone envision that such habits would eventually lead to large-scale deforestation of the continents—in Europe from 1100 to 1500, and in North America in the nineteenth and early twentieth centuries.

By 1800, Earth's population had grown to 1 billion people, some 250 times bigger than only 10,000 years earlier. The human population today is nearly 7 times bigger yet than in 1800, and with far greater technical capabilities. Because of human activities, ice, the force majeure of the planet only 20,000 years ago, is today in retreat, and perhaps on a trajectory to disappearance. No longer are humans passive adapters to the natural world—today we have become the principal agents of large-scale changes in the global environment.

CHAPTER 4
WARMING UP

Some people change their ways when they see the light; others when they feel the heat.

—CAROLINE SCHOEDER

The spreading of the last continental ice sheets over North America and Europe reached a maximum some twenty thousand years ago, and then the ice blanket began to melt back. The melting was the result of a warming of the climate over the next ten to twelve thousand years, an ameliorating change that brought Earth's average surface temperature upward by about fifteen to twenty Fahrenheit degrees. That change took Earth from the chill of the Last Glacial Maximum to a level a bit warmer than today, a warm thermal plateau called the Mid-Holocene Optimum. The ascent was not always smooth, as surges of fresh meltwater from the shrinking ice sheets spilled into the ocean, temporarily disrupting the currents transporting heat from the tropics to more remote and colder parts of the globe.

Reconstructing historic climate by reading the effects that a changing climate has on the natural world is both art and science, an endeavor known as paleoclimatology. It is not unlike piecing together a jigsaw puzzle, but with many pieces not quite fitting, and others missing altogether. And there is no picture on the box to guide you. The "pieces" that climate scientists work with come from both natural archives and human record-keeping, and are called climate proxies. A climate proxy substitutes imperfectly for a measuring instrument such as a thermometer or rain gauge. For a proxy to be useful in a historical reconstruction, it must also give an indication of the time when a climatic effect is being recorded.

For climate reconstructions we seek long records that encompass many years, and we prefer proxies that give year-by-year information. Trees turn out to be good proxies because they grow a little bit each year, and add a new ring of material around their trunks. The thickness of the ring indicates whether it has been a "good" or "bad" year for the tree. A thicker ring grows in a year when conditions are just right—not too hot, not too cold, not too wet, not too dry. In times of drought or thermal stress, growth is suppressed and the annual ring is thinner. Trees that are "on the edge," so to speak, at the very margins of the temperature range in which they can survive—in the polar latitudes or high up on mountains—are most sensitive to stress and therefore better proxies. And because the rings correspond to years, they can be counted and easily dated.

Snow accumulations that compress into ice are also good annual proxies, because there is usually a seasonal rhythm to snowfall that makes yearly accumulations appear as distinct layers in a glacier or polar ice sheet. Thicker layers, of course, indicate more snowfall, and the temperature at which the snow precipitated can be teased out of the oxygen and hydrogen isotopes in the ice's H_2O. In the marine environment, corals show an annual addition to their framework that reveals the temperature of the seawater in which the coral has been growing. Geographically well distributed proxies, over land and in the oceans, are necessary to reconstruct a global average temperature.

Wind-blown dust is an indicator of both aridity and wind patterns. As dust falls from the atmosphere it settles into the ocean, into lakes, and onto glacial ice. In all three settings it is incorporated into sedimentary or ice layers. The varying amount of dust in the layers indicates the changing susceptibility of Earth's land surface to wind erosion, which usually increases at times of drought. And the composition and mineralogy of the dust often indicate where the dust originated, so climatologists can reconstruct the pattern of atmospheric circulation from the dust source to the depositional site. Today satellite photos frequently reveal huge plumes of dust blowing westward off the Sahara Desert in North Africa—and this dust is slowly accumulating in sedimentary layers at the bottom of the Atlantic Ocean.

Human documents, such as records of agricultural production, human health, and sea ice extent, are also useful proxies. Just as trees grow better with enough sunshine and rainfall, so do agricultural crops. Year-by-year records of the wheat or maize yield, or of the grapes in vineyards, can serve as local proxies of weather conditions. Likewise, public health records can be massaged to reveal cold, damp years from warm, sunny ones. And for hundreds of years, European fishing fleets have kept good records of the geography of the sea ice that they encountered in the far North Atlantic.

So what do proxies tell us about the post-glacial climate? After reaching the thermal plateau some eight to ten thousand years ago, Earth slowly cooled about two Fahrenheit degrees on its way to the twentieth century. To be sure, there were some fluctuations superimposed on that slow descent, one known as the Medieval Warm Period, extending from about AD 950 to 1200, and a cooler event called the Little Ice Age, from about AD 1400 to 1850. But these were small departures in a long period of relative stability, a period that offered a golden opportunity for humanity to grow and spread. It was the time of the well-appointed, well-fed (and well-preserved) Iceman mentioned in chapter 2, who until the day of his demise was leading the good life of a central European who had already discovered the advantages of a varied diet and metal tools.

The Medieval Warm Period is best remembered as the time when Europeans first established settlements in Newfoundland and southern Greenland, complete with domestic animals that could graze on limited grass during the still short summer growing season. At the peak of settlement, Greenland hosted a few thousand people on a few hundred farms. These hardy settlers and their descendants remained in Greenland for a little more than three hundred years, until the climate shifted to a cooler phase that eliminated summer pasture and cereal crops and ended the European presence. The maximum temperature during the Medieval Warm Period has been reconstructed through proxy methods to be about at the temperature level of the mid-twentieth century—warm, but not as warm as early twenty-first-century temperatures.

The slow cooling from the peak of the Medieval Warm Period continued for some five hundred years and led directly into the Little Ice Age. In the North Atlantic region, the effects were well observed: The tenuous growing season in Greenland disappeared, and the seasonal sea ice extended much farther south into the Atlantic, making Iceland a much more difficult destination to reach. Alpine glaciers, rejuvenated with greater snowfall, advanced downward in their valleys, and crop failures became more common.

Both the Medieval Warm Period and the Little Ice Age are best documented in the North Atlantic region—in the Greenland ice cores, European tree rings, and the abundant historical records. But did these events mark only a regional change in climate, or were they global events, with effects seen elsewhere? In a sense the answer must be global, because even if the main expression of these climatic excursions occurred in the North Atlantic, the oceanic and atmospheric circulation would slowly "export" the effects to all parts of the globe, albeit in diminished scale and without simultaneity.

THE INSTRUMENTAL RECORD OF A CHANGING CLIMATE

In the early 1600s a major technological advance occurred that would transform our ability to monitor climate: the invention of the thermometer. This new technology put determinations of temperature changes on a quantitative basis, with an unprecedented uniformity and specificity. However, it was more than a century later, in 1724, that Daniel Fahrenheit, a German engineer, devised the scale of temperature that still bears his name, and two decades later, in 1744, when Anders Celsius, a Swedish astronomer, set forth his temperature scale. These two temperature scales survive to the present day, with 180 Fahrenheit degrees and 100 Celsius degrees separating the freezing and boiling points of water—thus each Fahrenheit degree is only five ninths as big as each Celsius degree. All countries that have adopted the metric system of weights and measures use the Celsius temperature scale. Of the world's major countries, only the United States continues to use the Fahrenheit scale.

The invention of the thermometer and the widespread adoption of either the Fahrenheit or Celsius temperature scale enabled temperature to be measured virtually anywhere and compared with measurements elsewhere. Thus the foundation for a network of calibrated instruments was laid, one that would provide the observations that show that Earth's average temperature, from the middle of the nineteenth century onward to the present day, has been rising.

At a few sites in Europe, such as the Klementinum Observatory in Prague, temperatures have been recorded daily for more than two hundred years. But from a global perspective, for a very long time after the invention of the thermometer, there were not enough measurement sites well distributed around the world to be able to make any global pronouncements. In particular, there were few measurements in the southern continents and vast regions of the oceans. Antarctica had no permanent meteorological station until 1903, when the Scottish National Antarctic Expedition established one in the South Orkney Islands. Since 1904 the station has been operated continuously by Argentina.

It was not just temperatures that were going unobserved in the seventeenth and eighteenth centuries—science in general was not exported to these regions for another two centuries. The religious brothers of the Society of Jesus, better known as the Jesuits, were an exception. They established meteorological stations and later seismographic observatories at several locations in Asia, South America, and North America. They appreciated that long historical records were necessary for the understanding of the environment in which they worked. But they were also early advocates of what today is called "liberation theology," and their attempts to educate indigenous peoples with such ideas did not sit well with the Spanish and Portuguese royalty and the Church authorities of the time. The Jesuits were recalled from their remote outposts and remained absent for two centuries, and the climate in many parts of the world went unobserved and undocumented.

At sea, save for the few meteorological stations established on islands, the record of sea surface temperatures had to be extracted from the logbooks of ships. Measurements of sea surface temperatures were made frequently aboard sailing ships by scooping up a bucket of seawater and sticking a thermometer in it. Later the measurement process shifted to the intake ports for seawater that was used to cool marine engines.

Today the taking of Earth's temperature occurs at thousands of locations around the globe, day in, day out, year in, year out. These observations of temperature take place on all the continents; at the sea surface by merchant, military, and scientific ships monitoring the temperature of the seawater as they traverse the global ocean; and by big arrays of fixed and floating buoys that automatically relay their temperature readings via satellite telemetry. The annual average of these many millions of individual thermometer readings around the globe is called the instrumental record of Earth's changing temperature. However, the measurement of temperature has had this global geographic coverage only over the past 150 years or so.

The merging of such large amounts of data, gathered in different ways at different times, requires careful attention to detail and the exercise of considerable quality control. Meteorologists sometimes change thermometers at weather stations, as new instrumentation is developed. They occasionally change the location of stations, as developing urban areas gradually surround formerly rural stations. The methods of measuring sea surface temperatures by ships at sea changed from buckets to water intake ports on ship hulls. And in doing the averaging, climatologists must be careful not to give too much weight to regions with many measurement sites as compared to other regions with far fewer sites. The good news is that the several organizations around the world that have independently developed methodologies to address these issues have produced very similar results.

THE VERDICT: EARTH IS WARMING

So what have the many millions of thermometer readings over some 150 years—the instrumental record—revealed? The fundamental result: they show that Earth's surface has on average warmed about 1.8 Fahrenheit degrees.

It has not been an unbroken climb for 150 years—there have been year-to-year ups and downs, some decades in which the temperature increased rapidly, and other decades when the warming slowed or was interrupted by some slight cooling. But one does not need to be a climate scientist to readily see that the graph of global average temperature over the past century and a half indicates a warming trend—a trend that

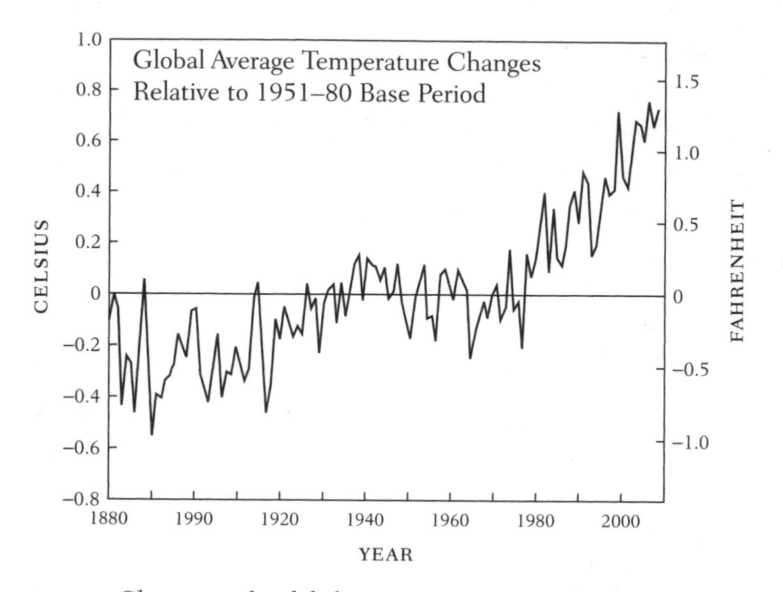

Changes in the global average temperature, shown as departures from the mean temperature over the years 1951–80. Data from the NASA Goddard Institute for Space Studies

is in fact accelerating. The warming trend over the past 25 years of the instrumental record is four times greater than that of the full 150 years. Earth's fever is rising rapidly.

In a geographical context, not every region displays the average behavior. Some parts of the globe have warmed more than average, some less, and a few areas have not warmed at all, or have even cooled. But the instrumental record is very clear: the average temperature of our planet's surface has increased significantly over the past 150 years. For the past half century it has been warmer than during the Medieval Warm Period. And during the last three decades, Earth's temperature has been rising faster than at any earlier time in the instrumental record.

Warming, moreover, is not confined to Earth's surface. Yes, the surface has warmed, but so has the lower atmosphere, the ocean water below the surface, and the rocks beneath the surface of the continents. Temperature measurements at depth within the oceans have been gathered from a number of different sources. Knowledge of the thermal structure of the oceans below the surface is important to submariners trying to cruise clandestinely and to fishing fleets seeking the habitats of favorite fish. Because sound travels faster in warmer water, an understanding of the temperature patterns in the ocean improves the accuracy of depth soundings—determinations of water depth based on how long it takes a pulse of sound to travel from the surface down to the seafloor and bounce back to the surface.

Much of this temperature data, collected over many decades, now resides in the U.S. National Oceanographic Data Center, and has proved to be a treasure trove of information about how the oceans have responded to the warming that has taken place at the surface. These data show that since about 1950 the oceans have been absorbing heat at a measurable rate, to depths of about 10,000 feet, with about two thirds of the heat stored in the upper 2,500 feet.[1]

The temperature of the rocks beneath the surface of the continents also shows the effects of a changing climate. Much of my own scientific work over the past two decades has been devoted to collecting and analyzing subsurface temperatures from around the world, to reconstruct the climate history the rocks have experienced. The principles behind this geothermal method are straightforward: a rock placed next to a campfire in the evening will still be warm in its interior in the morning, long after the campfire has burned out. The warm temperatures in the interior can, with the help of a little mathematics, reveal when, how long, and how hot last night's campfire burned—in other words, it can reveal the "climate history" that the rock's surface was exposed to the previous evening.

In the Earth, we measure temperatures at intervals down deep boreholes that have been drilled into the rocks of Earth's crust, or into the thick ice sheets of Greenland and Antarctica. The rock holes, penetrating

1. S. Levitus et al., "Warming of the World Ocean," *Science* 287 (2000): 2225–29.

to depths of 1,000 to 2,000 feet, have usually been drilled in search of minerals or water, or, in a few cases, for scientific research. The profiles of temperature down the holes reveal depth ranges over which the temperatures are either higher or lower than the temperature expected at that depth in the absence of any changes in the climate. These anomalous zones are the remnant signatures of past temperature fluctuations at the surface that have propagated downward into the subsurface. The Little Ice Age can be "seen" in the temperatures 500 feet down in the Greenland ice sheet, and the warm plateau of the mid-Holocene at depths between 1,500 and 2,500 feet.

How long can rocks and ice "remember" their thermal history? The pace at which heat is transferred through these materials is very slow, so slow that any fluctuations of surface temperature, either increases or decreases, since the last glacial maximum 20,000 years ago will have propagated no deeper than a mile or two into the subsurface. So the uppermost part of Earth's continental crust is effectively a thermal archive of climate change over the past several thousand years.

My colleagues and I have studied more than eight hundred borehole temperature records from around the world in some detail, and have been able to show that five centuries ago, Earth's average temperature was about two Fahrenheit degrees (a little more than one Celsius degree) lower than today.[2,3] Since the year 1500, Earth's rocks have warmed, slowly at first and more rapidly later—fully half of the warming occurred in the twentieth century alone. And the surface temperature changes interpreted from the rocks are fully consistent with, but totally independent of, the instrumental record on the continents during the period of overlap, 1860 to the present. Scientific conclusions are always

---

2. S. Huang, H. N. Pollack, and P.-Y. Shen, "Temperature Trends over the Past Five Centuries Reconstructed from Borehole Temperatures," *Nature* 403 (2000): 756–58.
3. H. N. Pollack and J. Smerdon, "Borehole Climate Reconstructions: Spatial Structure and Hemispheric Averages," *Journal of Geophysical Research* 109 (2004): D11106, doi: 10.1029/2003JD004163.

more persuasive when the same conclusion is reached by more than one independent method.

Just as the water mass of the oceans and the rocks of the continental crust are warming, so also is the atmosphere above the surface. Temperatures taken by instruments aboard weather balloons since the 1950s and from orbiting satellites since the late 1970s have provided a picture of the temperature trends at various levels in the atmosphere, albeit over a considerably shorter time than is available with the instrumental record at the surface. And the task of assessing the temperature from a satellite looking down at its target from above, rather than being immersed in it, is not an easy one. Initially there was a suggestion that the satellite record was at odds with the measurements at the surface, but as the technical difficulties of the satellite measurements were recognized and resolved one by one, the differences largely disappeared. Today these two independent estimates of the surface temperature trends are very similar.

The measurements of temperature with scientific instruments the world over—at Earth's land and sea surface, in the deeper waters of the ocean, in the rocks of the continents, and in the thin atmospheric envelope above the surface—all are telling the same story: planet Earth is, without question, warming.

THE TRENCHES OF DENIAL

In the decades sandwiching the end of the twentieth century and the beginning of the twenty-first, probably no other scientific topic was more in the news, and more contentious, than Earth's changing climate. It achieved prominent coverage and editorial commentary in the *New York Times*, the *Washington Post*, and the *Wall Street Journal*, as well as cover story status in *Time*, *Newsweek*, *The Economist*, *BusinessWeek*, *Vanity Fair*, *The Atlantic Monthly*, *Skeptical Enquirer*, *New Scientist*, *Scientific*

American, Wired, Sports Illustrated, and many other magazines. Former vice-president Al Gore produced his film *An Inconvenient Truth*, and the Weather Channel has its weekly *Forecast Earth*. Both the U.S. Senate and House of Representatives have held formal climate change hearings.

Initially the media presented climate change as a "he said, she said" story, with little analysis of conflicting positions. It became apparent in this "fair and balanced" coverage that there was a not-so-subtle subterfuge taking place, in which prominent players in the carbon-based energy industry[4] (e.g., Peabody Coal and ExxonMobil) had invested quietly to interject misinformation and uncertainty about climate science into the discussion, describing it with terms such as "unsettled science" and "uncertain science," or more boldly attempting to discredit the accumulating scientific results as "unsound science" or "junk science." A handful of contrarian scientists—many who were financially supported by the fossil fuels industry—took issue with the emerging climate science consensus and became known as skeptics or climate contrarians.

Although the attacks on the developing scientific consensus about terrestrial climate change seemed in some ways to come from a shotgun, the essential position of denial could be distilled into four main elements:

1. The instrumental record of surface temperature change was flawed.
2. The causes of climate change were entirely natural.
3. The consequences of climate change would be beneficial.
4. The economic cost of addressing climate change would not be worth the effort.

These four elements were in effect sequential trenches of defense occupied by those ideologically opposed to the concept of anthropo-

4. See Ross Gelbspan's *Boiling Point: How Politicians, Big Oil and Coal, Journalists, and Activists Are Fueling the Climate Crisis—And What We Can Do to Avert Disaster* (New York: Basic Books, 2004).

genic climate change, or blind to its reality. If any one of these assertions could be persuasively demonstrated or proven, the rest of the list would be rendered irrelevant. The defensive arsenal was (and continues to be) well stocked with misinformation, irrelevancies, half-truths, misunderstandings, oversimplifications, and outright falsehoods, but underlying all was a notion of seriality: the climate contras viewed the entire climate change argument as a long chain of evidence, and if any link could be broken, then the chain could no longer carry any weight and the climate change concept would fall apart. In reality, the scientific story of climate change is much more like a net hammock of interwoven strands of evidence—if one strand proves weak, there remain many that continue to support the growing reality of the climate change saga.

The climate contras recognized that if the first trench could be successfully defended—if they could make the case that there were no compelling observations of a changing climate—the war would effectively be over. So, they rolled out mortars that lobbed argument after argument to a puzzled and largely scientifically illiterate public, attacking the instrumental record of a warming Earth:

- "We shouldn't be placing much credence in data from weather stations in cities, because the 'urban heat island' effect is contaminating the record."
- "You can't trust a century-long record of thermometer readings when they change thermometers every couple of decades."
- "How can climatologists argue that Earth is warming when we here in Graniteburg, New Hampshire, are experiencing the worst winter in memory?"
- "How could anyone say the globe is warming when for the past umpteen years, Dry Valley Crossing, Nevada, has been cooling?"
- "Satellites taking Earth's temperature don't show any warming."
- "Maybe the continents are warming, but the oceans are cooling."

These arguments, sometimes raising interesting scientific or technical questions, have all been addressed: The urban heat island effect is real, but it has been corrected for, or sidestepped by using only rural meteorological data on land. And one must remember that there are no cities sitting on 70 percent of Earth's surface—the oceans. The effect of changing thermometers can be assessed by using both the new and old thermometers side by side for a time to be sure they give the same results. Whether someplace is having a very cold winter, or a very hot summer, is an irrelevancy—climate change is about long-term trends in temperature, not about year-to-year oscillations. That someplace may actually be cooling when the globe on average is warming is also irrelevant—as noted earlier, some places may be warming more than the average, some less than average, and a few might actually be cooling, but the overall average remains one of warming. It would be a rare climate system where every place did exactly the same thing. I have already mentioned that the early apparent differences between satellite and surface measurements have now been resolved. And are the oceans cooling while the continents are warming? Hardly. The long-term trend of temperature in all the oceans of the world is a very consistent warming for the past half century.

NATURE'S OWN THERMOMETERS

Amidst the lunges and parries about the accuracy of the instrumental record, it is easy to lose sight of the fact that one need not rely at all on scientific instruments to make a persuasive case that Earth's climate is warming. Nature has her own thermometers—plants and animals that inhabit the land and the sea. Flowering plants that take their annual cues from the warming and cooling of the seasons are now sprouting and blooming earlier in the spring, and birds are laying their eggs earlier. Birds that time their annual migrations by changes in temperature are lingering longer in the fall before departing for their winter habitat—and some are no longer migrating at all because winters have become so mild. Insects have begun to migrate up mountains as the warming adds new terrain to their ecosystems, and some insect population dynamics have suddenly changed when mild winters no longer are cold enough to kill off most of the previous summer's residents. And as lake waters have warmed, their fish populations also change—coldwater species such as walleye and trout are being gradually replaced by warmer water bass and bluegills.

The timing of natural events is a part of the biological sciences called phenology, and observing the timing of seasonal arrivals and departures, of blooming and folding, of hatching and fledging, has long been a favorite activity of amateur naturalists. The routine collection of phenological and environmental data such as temperature and precipitation over long time intervals is vitally important to understanding the behavior of the climate system. But this type of scientific work is not glamorous. It often is done by unheralded people—some professional, some amateur—who receive no substantial reward or recognition save for the knowledge that they are contributing to a body of data that ultimately has immense scientific value. Euan Nisbet, an atmospheric scientist at Royal Holloway College of the University of London, has commented that "monitoring is science's Cinderella, unloved and poorly paid."[5] Let me describe some of this Cinderella science.

At the Mohonk Mountain House, a resort some eighty miles north of New York City, a meteorological observing station sits atop a rugged outcrop of rock. Since 1896 someone has trudged up the outcrop every day to read the thermometers and rain and snow gauges that are housed there.[6] Over the full more-than-a-century period of observation,

5. Euan Nesbit, "Cinderella Science," *Nature* 450 (2007): 789–90.
6. Anthony DePalma describes the Mohonk Mountain House weather station and the dedicated observing corps in an article in the *New York Times*, September 16, 2008, D1.

the "someone," in fact, has been only five individuals, with Daniel Smiley, Jr., a descendant of the founders of the resort, doing the duty for a half century. He made notes of many other phenomena, such as the first blooming of this or that flower each year, the first arrival of various birds in the spring, and the temperature and acidity of a nearby small lake, thus compiling a remarkable record of natural history and change at this location. At Mohonk Mountain, these thousands of daily observations show that since 1896 the average annual temperature has risen 2.7 Fahrenheit degrees and the growing season has been extended by ten days.

ON A RESEARCH TRIP to Russia in 2001, I spent several days in Irkutsk, situated in southern Siberia, just north of the border with Mongolia. Irkutsk is just about as far to the east of the Greenwich prime meridian as my home state of Nebraska is west. It is a stop on the Trans-Siberian Railway, but it is not a new railway city like Novosibirsk, a thousand miles west along the line of rail. In 1727, Vitus Bering, on his three-year overland trek on horseback to Kamchatka, to begin his voyage of discovery of the Bering Strait, wintered in Irkutsk because of its "amenities." Irkutsk remains a place of stark contrasts—three-hundred-year-old small ornate wooden houses juxtaposed with Soviet-style concrete apartment blocks. Less than an hour from Irkutsk is the foot of Lake Baikal, the oldest and deepest lake with the largest volume of freshwater in the world. It is a narrow lake, but almost four hundred miles long and a mile deep, occupying a tectonic rift valley, similar to Lakes Tanganyika and Malawi in East Africa.

When I went to Lake Baikal in early April of 2001, it was still frozen tight with an ice lid three feet thick—the spring ice breakup was still six weeks away. In conversations with scientists at a solar observatory overlooking Lake Baikal, I learned of a remarkable family at a small biological research station just a few miles away. This family had been studying Lake Baikal for three generations. Mikhail M. Kozhov had come to Irkutsk State University shortly after the end of World War II, and began making temperature measurements and biological surveys in the waters of Lake Baikal.[7] In summer he worked from a boat; in winter, through holes drilled through the ice. His daughter, Olga Kozhova, assisted him, and when he died in 1968, she continued the program of measurements, later assisted by her daughter, Lyubov Izmesteva. Olga died in 2000, but Lyubov, herself now a professor at Irkutsk State University, continues the measurements. This archive of temperature data shows that the surface waters of Lake Baikal have been warming at a rate of about one Fahrenheit degree every twenty-five years, and the warming is slowly penetrating to greater depths.

COOPER ISLAND IS a small low-lying barrier island a short distance off Point Barrow, Alaska, the northernmost point of the United States and indeed of North America. It sits some three hundred miles north of the Arctic Circle and is the nesting and breeding site of a colony of black guillemots, a not-so-common seabird of the Arctic. In 1972, a young ornithologist named George Divoky began a study of the breeding habits of these birds.[8] Every summer, for the next thirty years, he spent on Cooper Island with the black guillemots and an occasional polar bear. Most summers were in "solitary confinement," but occasionally he took a field assistant. Carefully he noted the dates when the guillemots returned to Cooper, the dates when they laid their eggs, the dates when chicks hatched and later fledged. What he has discovered is that the entire reproductive sequence has shifted more than ten days earlier in the Arctic summer. Guillemots will nest as soon as the snow

---

7. Cornelia Dean describes this three-generation family endeavor in her *New York Times* article of May 6, 2008.
8. Darcy Frey describes George Divoky's three-plus decades of research on Cooper Island in an article in the *New York Times Magazine* of January 6, 2002.

melts, but no sooner, and so the earlier nesting of these seabirds serves as a proxy for the timing of the annual snowmelt. But there was also some bad news for the guillemots: their population began to diminish around 1990. Year by year the Arctic warming has been moving the sea ice much farther from the nests on Cooper Island. Because the margins of the sea ice are favorite feeding spots for these seabirds, the retreat of the sea ice has been slowly moving food almost out of their reach.

NATURE'S BEST THERMOMETER, perhaps its most sensitive and unambiguous indicator of climate change, is ice. When ice gets sufficiently warm, it melts. Ice asks no questions, presents no arguments, reads no newspapers, listens to no debates. It is not burdened by ideology and carries no political baggage as it crosses the threshold from solid to liquid. It just melts.

THE ICE SEASON

When I was a boy growing up in eastern Nebraska, the calendar of certain activities was set by the seasons. Neighborhood hockey started up when nearby George's Lake froze over, and duck hunting began as the myriad channels of the Platte River became choked with ice. The family springtime fishing trip to the boundary waters of Minnesota was determined by the breakup of the winter ice six hundred miles to our north. In late May my father would be on the phone to friends in International Falls, Minnesota, checking whether the ice had moved out, and one week after the breakup, we were there trolling for walleyes. The rhythms of communities the world over have similarly been tied to the comings and goings of the annual ice.

Madison is the Wisconsin state capital and home to the University of Wisconsin. The city sits between Lakes Mendota and Monona, bodies of water that have provided recreational activities for residents ever since the city was founded in 1836, the same year the Wisconsin Territory was created. Perhaps not surprisingly, the dates of fall freezing and spring breakup have been dutifully recorded for almost a century and a half, and they tell a very interesting story. In 1850, Lake Mendota froze in early December and broke up in early April, but 150 years later, freezing had shifted to some nine days later and breakup occurred almost two weeks earlier.

Along the eastern shore of Lake Michigan is Grand Traverse Bay, another location with diligent record-keepers since the mid-nineteenth century. The long record compiled there shows that since 1851 the bay froze over completely at least seven times in *each* of the first twelve decades; this figure dropped to six times in the 1980s, three times in the 1990s, and only twice in the first decade of the twenty-first century.[9] For the years when the bay has frozen over, the period of winter ice cover has diminished by thirty-five days.

But it is not just lakes in North America that are showing trends toward shorter intervals of annual ice cover—in Scandinavia and Europe, in Asia and Japan, the long-term observations are telling the same story. And it is happening not just to lakes—major rivers leading to the Arctic Ocean, such as the Mackenzie in Canada and the Angara and Lena in Siberia, show similar trends. Wintertime ice in the freshwater of the Northern Hemisphere is becoming a much rarer commodity.[10]

Mountain glaciers everywhere—in New Zealand, the Andes, the Alps, Alaska, the Rocky Mountains, Central Asia, equatorial Africa—shrank

9. Jim Nugent, a horticulturalist in the Michigan State University Extension Service, kindly provided the Grand Traverse Bay freezing statistics.
10. J. J. Magnuson et al., "Historical Trends in Lake and River Ice Cover in the Northern Hemisphere," *Science* 289 (2000): 1743–46.

over the twentieth century. The U.S. Geological Survey and the U.S. National Snow and Ice Data Center have collected air and ground photography of glaciers in the United States showing the extent of glaciers at various times in the past.[11] In Glacier National Park in Montana, the melt-off has been dramatic—of the 150 glaciers present in 1850, fewer than 30 are still present today. At the present rate of melting, none will survive past 2030.

Mount Kilimanjaro sits just a few degrees south of the equator, in East Africa. The equator is an unlikely place to find natural ice, unless you go very high. Kilimanjaro reaches more than nineteen thousand feet above sea level, and for as long as anyone can remember it has had snow and ice at its peak. This iconic image of Africa was immortalized in Ernest Hemingway's short story "The Snows of Kilimanjaro." But throughout the twentieth century, Kilimanjaro has lost ice steadily. The volume of ice present in 2008 was less than 10 percent of what it was a century ago—and at the present rate of loss, ice will disappear from equatorial Africa by 2020.

The Athabasca Glacier in the Canadian Rockies of Alberta is perhaps the most visited glacier in North America, by virtue of its position between Banff and Jasper national parks, two of Canada's favorite scenic treasures. The recession of this glacier is well marked by a succession of signposts installed over the years at the snout of the glacier, which show the glacier's extent at various times in the past. Over the past 125 years the Athabasca has receded almost a mile from the first signpost.

The snowfields and glaciers of the European Alps are also shrinking rapidly, so rapidly in fact that the tourist industry is resorting to desperate measures to slow summertime melting, including laying reflective sheets over the glaciers, as a giant seasonal sunscreen. At the present

11. http://nrmsc.usgs.gov/repeatphoto/. http://nsidc.org/data/glacier_photo/repeat_photography.html.

rate of melting, Alpine glaciers will be only memories by the end of this century. In Asia the glaciers in the Himalayas each year are losing ice equivalent to the entire annual flow of the Huang He, China's fabled Yellow River.

TUNDRA TRAVEL DAYS

Getting around in the Arctic terrain is never simple, but it is easier, ironically, in winter than in summer. To be sure, the unending daylight of summer offers visibility of unimagined scale, and ease of navigation in an area with few human landmarks. But the broad vistas disguise the fact that in the summer, the ground becomes soft and spongy, depriving vehicles of a firm surface to traverse. The permafrost, the terrain that experiences an average annual temperature below the freezing point, undergoes some limited summertime melting in what is known as the "active zone." This zone extends downward a foot or two or three, and turns a frozen-hard wintertime surface into summertime mush. Off-road traffic (and there are very few roads) becomes impossible. Thus, overland transport of supplies to mineral and petroleum exploration camps, scientific stations, and remote settlements is, of necessity, confined mostly to winter.

The time of year when such transport can take place is known as the tundra travel season, and is measured in terms of the number of days that vehicle passage overland is possible. Tundra travel days are rapidly diminishing in number. In 1970 one could roll over the frozen surface of northern Alaska more than seven months of the year, but today such travel is possible during only four months, from early January until about mid-May. The overland travel window is closing at a rate of about one month per decade. The tundra surface is now an "active zone" two thirds of the year, and in another half century it may be impassable year-round.

GREENLAND

Greenland is an Arctic island bigger than Mexico. It sits almost completely north of the Arctic Circle on the North American side of the Atlantic Ocean. It is a huge reservoir of ice, in volume second only to that of Antarctica. The ice on this large island, covering all but its coastal fringe, is equivalent to more than twenty feet of sea-level change, were it to return to the sea. The top of the ice pile is about twelve thousand feet above sea level, with another thousand feet below sea level because the ice load has depressed the rocky surface beneath it. Greenland's ice slowly creeps downward, and spills into the sea in hundreds of glacial streams around the periphery of the island.

The glaciers are like small holes around the base of a rain barrel—some water escapes through each hole, and in the absence of precipitation, the water level in the barrel will slowly decline. When precipitation into the barrel equals the water losses through the holes, the water level in the barrel remains unchanged, and when the rainfall exceeds the losses at the bottom, the water level will go up. Were there no replenishment of ice in the interior from snowfall, Greenland would eventually be drained of ice.

Every year Greenland undergoes summertime melting around the perimeter of the ice sheet, where the seasonal temperatures at low elevations are sufficiently warm. This band of melting on the fringes has been more or less stable in areal extent and in elevation throughout most of the twentieth century, but toward the end of the century the zone of melting began to creep to higher elevations and over larger areas. The fraction of Greenland's area that undergoes summer melting is 30 percent greater today than it was only thirty years ago, and now ice melts at elevations greater than six thousand feet above sea level.

In midsummer the melting areas are dotted with meltwater pools and lakes, beautiful blue jewels accenting the white backdrop. Some of these bodies of water will refreeze in winter, and thus do not represent a net loss of ice mass. Others lose their water in streams that run to the sea, and these do represent a net loss of ice mass and contribute to a rising sea level.

It is not an easy task to determine whether the ice budget of Greenland or Antarctica is in surplus or deficit. One technique that has been employed is called repeat-pass airborne laser altimetry, a method in which an aircraft flies over the ice surface at a low but steady altitude and repeatedly flashes a laser beam at the surface below. The beam is reflected from the surface back to the aircraft. The time it takes for the laser beam to go down to the ice and return can be translated into the elevation of the ice surface.[12] A repeat of the measurement in a few months will show what changes have taken place in the elevation of the ice surface. If the surface has become lower, there is a deficit, and if it is higher, there has been an accumulation. But elevation changes can be misleading—a fresh snowfall might add three feet in elevation, but because the new snow is light and fluffy, it doesn't represent the same mass as three feet of dense glacial ice.

Another technique to determine if an ice budget is changing makes use of detailed measurements of Earth's gravity, as felt by scientific satellites as they orbit the planet. Very small changes in gravity are associated with the different densities of the various rocks that make up Earth's crust. Compared to the average, a low-density rock has a mass "deficiency" and a high-density rock has a mass "excess." The force of gravity increases over areas of excess mass and decreases over regions of mass deficiency. The paths of Earth-orbiting satellites are perturbed very slightly—sped up or slowed down a tad—by these small variations in local mass and gravity. Thus careful observations of satellite orbits

12. This technique is analogous to the determination of ocean depths using sound transmission.

can over time reveal whether a region is losing or gaining ice. A special satellite experiment known as GRACE (Gravity Recovery and Climate Experiment) has been operating since 2002, paying special attention to Greenland and Antarctica. In Greenland, GRACE determined that there is an ongoing ice mass loss tied to an acceleration of the glaciers draining the interior. The ice deficit for Antarctica has also increased, by 75 percent over the past several years, principally because of accelerating glacial flow following the disintegration of floating ice shelves around the continent.

AN OCEAN OF ICE

The Arctic Ocean is a roughly circular ocean with the geographic North Pole at its center. The diameter of the ocean is about 2,800 miles, with North America and Greenland sitting on one side, and Europe and Asia on the other. The entire ocean lies north of the Arctic Circle, and thus experiences the annual extremes of solar illumination—including some days of around-the-clock darkness in winter and unending daylight in summer. For as long as people have been paying attention, much of the ocean has remained frozen year-round in a vast sheet of sea ice. In summer, some of the sea ice breaks up and melts to expose open water, but in winter it refreezes, in a layer about three to six feet thick. During the first half of the twentieth century, about one third of the sea ice melted and refroze each year, leaving two thirds of the ocean with older ice, up to about five years old in places. The older ice is also thicker, occasionally reaching a thickness of fifteen feet or more. Even though much of the Arctic Ocean has been covered with sea ice for at least as long as humans have observed it, the ice is not the same ice. Because sea ice is always on the move—drifting from the Far East, over the North Pole, on toward Scandinavia, and exiting into the Atlantic—no extensive region of the Arctic Ocean has ice much older than five years. The exceptions are in the narrow channels that surround the many islands of the Canadian Arctic, outside of the mainstream of the Arctic drift.

The Age of Exploration—roughly the sixteenth through the nineteenth centuries—coincided with the Little Ice Age cool interval. In the Arctic Ocean, ice formed in every nook and cranny, including in the many channels that wind their way through the archipelago of islands comprising the northern territory of Canada. This maze of waterways, were they to become ice-free, would allow a maritime shortcut from Europe to the trading nations of Asia, a route shorter by two thirds compared to the alternative routes around either Africa or South America. This passage, more concept than reality, was called the Northwest Passage.

For most of maritime history, however, this route has been closed with ice. Time and again the ice thwarted attempts to open this new trade route to the Orient. On his third and last voyage of discovery aboard HMS *Discovery*, Captain James Cook searched for the western entry to the passage. Sailing west along the Aleutian Islands in the summer of 1778, he crossed into the Bering Sea near Unalaska Island, and then along the western coast of Alaska to the Bering Strait, with still no hint of a pathway to the east. Northward he continued, through the Bering Strait to latitude 70° north, where the land began to ease off to the east. Was this the western portal? Cook excitedly began the mental calculations of how long it might take to reach Baffin Bay, that stretch of open water between Canada and Greenland some two thousand miles to the east.

But it was not to be. Two days later Cook saw ice blink, the reflection of a vast expanse of ice on the low clouds in the distance. In a few hours the ice came into view, a solid wall more than ten feet high, as far as the eye could see. Cook recognized the futility of continuing, and turned around to retrace his course back into the Pacific. It would be his last glimpse of ice ever—six months later Captain Cook was dead, killed in a battle with native Polynesians in Hawaii.

Others tried to navigate the Northwest Passage from east to west with no more success. The storied Franklin Expedition of 1845–47 became ice-locked about midway through the passage, and all aboard perished from starvation. It was not until 1906, when Roald Amundsen, the Norwegian explorer who would five years later gain fame as the first person to reach the South Pole, completed a three-year journey through the Northwest Passage to reach Alaska. But the ice he faced may already have been less of an obstacle than that encountered by the eighteenth- and nineteenth-century explorers—the Little Ice Age had reached its peak in the nineteenth century. By 1906, Amundsen was already benefiting from a warming climate.

The summertime retreat and winter refreezing of sea ice are regular cyclical occurrences of long standing, but in the latter decades of the twentieth century, the summer melting began to consume much more than the usual amount of ice, and the winter refreezing fell short of restoring it. By the end of the twentieth century, the summer sea ice had diminished by some 25 percent from its mid-century extent. And as the older ice was replaced by younger ice, the average thickness of the sea ice also diminished, to about half its mid-century measurement.

The Russian icebreaker *Yamal* for a number of years has ferried tourists to the North Pole, for a "picnic" on the ice. In August of 2000, everyone aboard was in for a surprise—when *Yamal* reached the pole, there was only open water. Occasional open water in sea ice is not uncommon—such an ice-free area is called a polynya, a Russian word now in the international lexicon. Polynyas come and go, vagaries of upwelling ocean currents beneath the ice and the wind above. A few polynyas are more or less permanent geographic features, reflecting the stability of the ocean currents, but many others are transient—here today, gone tomorrow. We do not know how common or how rare a North Pole polynya may be, but those who witnessed the occurrence in 2000 remarked that the sea ice had been very thin and peppered with polynyas all the way to the pole. *Yamal*'s captain said that in all the years he had been traveling to the pole, a polynya there was a first for him.

THE NORTHERN HIGH latitudes are not unique in signaling a changing climate. In the south, all around Antarctica, the ice is also growing restless. Big ice is the norm in Antarctica, and after coming to the white continent for eighteen years, my jaw does not drop easily. Yet in mid-December 2007, I was awestruck with what was unfolding on the horizon. We were at latitude 61° south, longitude 54° west, between Elephant and Clarence islands, at the tip of the Antarctic Peninsula, when the biggest piece of floating ice I had ever seen came into view. Thirty-one miles long, twelve miles wide, edged with sheer ice cliffs reaching more than a hundred feet above the sea surface, and another eight or nine hundred feet below—a massive island of ice adrift in the southern ocean.

This great slab may have broken away from the Filchner Ice Shelf deep in the Weddell Sea, or perhaps was a fragment of an even bigger mass that had separated from the Ross Ice Shelf some 2,500 miles south of New Zealand in 2001. Numbers cannot fully describe this floating behemoth. Sixty cubic miles of ice? Fifteen times the area of Manhattan? The volume of water in Lake Erie? There it was, gigantic ice adrift on a journey to nowhere, pushed along by wind and currents at about two miles per hour.

Sailing alongside this floating ice island I found it impossible to capture the scale—a photo simply showed a cliff extending from the foreground to the horizon. One needs to step back—way back—to be able to see this slab in its entirety. Actually, one needs to step up about a hundred miles, to the viewpoint of an Earth-orbiting satellite, to capture this slab in a single frame. The experience is not unlike feeling the fierce wind of Hurricane Katrina on the ground in southern Louisiana, but needing a satellite image of a giant spinning pinwheel covering the entire Gulf of Mexico to see the full scale of nature's atmospheric fury.

The breaking away of ice of this magnitude from the outer edge of an Antarctic ice shelf certainly begs for attention, particularly when it is not an isolated phenomenon. The Ross Ice Shelf, about the size of France, is the biggest of the huge floating ice sheets nestling along the margins of Antarctica. Others abut both sides of the Antarctic Peninsula, the long and narrow finger-like mountain chain that stretches toward South America. Along the peninsula, mountain glaciers drain ice from the high places, sending it to the sea, where it floats in giant sheets that extend tens and hundreds of miles away from the rocky coast. The Larsen, the Filchner, the Ronne, and the Wilkins ice shelves—named for whalers, scientists, and explorers of a century ago—also are showing wear and tear today.

In early 1996, when I was working aboard MS *Explorer* (the expedition ship that sank not far from Elephant Island a decade later), the captain and expedition leader revealed to the expedition staff that we were going to attempt the first-ever circumnavigation of James Ross Island, named for Sir James Clark Ross, a British explorer who navigated the region in 1842 (and for whom the Ross Ice Shelf is also named). James Ross Island lies near the tip of the peninsula, on the east side; it is the eleventh largest of the myriad islands that dot the fringe of

Antarctica. It had been bound tightly by ice at least since any human had viewed it. Yet, there were hints that a circumnavigation might be possible.

Just to the south lay the Larsen Ice Shelf, one of the big ice shelves attached to the peninsula. Two years earlier, the northernmost segment of the Larsen, an area about the size of Luxembourg, had disintegrated, flushing great icebergs into the adjacent Weddell Sea. Could the icy handcuff holding James Ross Island also be loosening? We thought it was a possibility, and began our push into the ice. We were not to be rewarded, however, because the ice was not ready to yield its grip—but the very next year, *Explorer* succeeded making it around James Ross Island, through channels that had not seen open water for thousands of years.

Only five years later an even larger segment of the Larsen disintegrated, one as large as Rhode Island, in a spectacular one-month breakup that delivered so much floating ice to the region that ship navigation was substantially impeded and only ships with a scientific mission ventured into the area. And in late March of 2008 the Wilkins Ice Shelf, on the southwest side of the peninsula, an area about half the size of Scotland, began to disintegrate, shedding floating ice islands of Brobdingnagian scale into the sea. The initial "sliver" from the edge of the shelf was 25 miles long and 1 mile wide, and once it was separated, another 150 square miles behind it quickly broke up. New fractures in the remaining shelf appeared in a November 2008 photo, indicating that the breakup was still progressing, and by April 2009 the disintegration was complete.

The ice shelves around Antarctica are great sheets of ice that have ponded up around the mouths of glaciers that drain ice from the interior. Parts of the shelves may be grounded, but much of the ice is floating as large sheets on the sea. These massive, partially anchored shelves serve as buttresses that slow the outflow of the glaciers that nourish them—but when the shelves disintegrate, the glaciers find new freedom and speed up their delivery of ice to the sea. A recent survey[13] of all the outlet glaciers around Antarctica shows little net loss of ice from East Antarctica (the bigger fraction of Antarctica, east of the Transantarctic Mountains), but substantial and increasing ice loss from West Antarctica and the Antarctic Peninsula.

WHAT DOES THIS accelerating ice loss from both Greenland and Antarctica mean? In a bathtub, the volume of water determines how high the water reaches, and the same holds true for Earth's great natural bathtub—the ocean basins. The volume of water in the oceans rises and falls during the comings and goings of ice ages, and these hydrological transfers are accompanied by changes in sea level of several hundred feet. But in the period of general warmth and relative stability that we have experienced over the past ten thousand years, we have not seen dramatic changes in sea level. There has been a rough equilibrium between losses from the oceans through evaporation, and returns to the oceans via precipitation and the flow of rivers and glaciers to the sea. Over the past several millennia these withdrawals and deposits have continued to take place in the oceanic account, but the balance has remained pretty steady.

But the twentieth-century warming of Earth and the loss of ice from the continents are beginning to change the oceanic balance—in the upward direction. Two principal factors are at work. The first is the increased melting of ice and the return of the meltwater to the sea, or alternatively the direct deposit of ice into the sea from faster-flowing glaciers. The second factor is the volumetric expansion of seawater as the oceans warm. The volume of most liquids increases when the tempera-

13. E. Rignot et al., "Recent Antarctic Ice Mass Loss from Radar Interferometry and Regional Climate Modeling," *Nature Geoscience* 1 (2008): 106–10.

ture goes up—that is the fundamental principle behind liquid-in-glass thermometers that show the level of the liquid rising in the scaled glass tube as the temperature rises.

Changes of water level during a flood episode along a river are apparent in vertical changes—how high the water rises along a levee or the wall of a building—and in horizontal changes seen in the growth in the area covered by water. Sea-level changes are apparent in these same ways. The measurement of the vertical change, the amount of sea-level rise, is accomplished by an instrument called a tide gauge, the primary purpose of which is to measure the amplitude of the high and low tides in a bay or harbor. This instrument records the changing water level during the daily rise and fall of the tide, but over years, decades, and centuries it also shows the long-term changes associated with slowly rising sea level. Data from thousands of tide gauges the world over have been collected and analyzed, and show that during the twentieth century, sea level on Earth rose about eight inches. One third of the rise comes from new deposits of meltwater and ice into the sea, and two thirds from the thermal expansion of the warming oceans.

But it is the horizontal incursion of the rising sea that is most apparent to the eye. On a gently sloping beach, a small rise of sea level will extend quite a ways inland to define a new shoreline. Eight inches of sea-level rise on a beach with a gentle slope of one degree will move the shoreline almost forty feet inland. And the same forty feet will be subject to the daily flooding of the high tide, and more vulnerable to the high-water surges from storms at sea.

THE IPCC

In 1988 the United Nations created an international scientific working group called the Intergovernmental Panel on Climate Change (IPCC). The charge to this group was to assess whether climate change was occurring, what the causes of such climate change might be, what the consequences of past and future climate change have been and might be, and what options might exist for mitigation of, and/or adaptation to, a changing climate on Earth. The task of organizing this panel fell to the World Meteorological Organization and the United Nations Environment Programme, both entities of the United Nations.

The IPCC is not a research organization itself, but rather an evaluator and summarizer of peer-reviewed scientific research published in scholarly journals the world over. It issues periodic assessment reports every five years or so, describing the state of knowledge about climate change. These assessment reports have appeared in 1990, 1995, 2001, and, most recently, 2007. Components of the reports are written by teams of scientists active in the various subfields of climate science, then collected into chapters by a group of "lead authors"; the chapters are assembled into a coherent and seamless assessment by an even smaller number of "coordinating lead authors." Altogether, more than two thousand active climate scientists contributed to the Fourth Assessment Report published in 2007.

A few words about the process of peer review in the IPCC assessments: peer review is essentially a process of quality control in the world of serious scholarship. Publication of research results in a peer-reviewed journal means that an article has been read by other practicing researchers in the area, and assessed for originality, appropriate methodology, data quality, and sound conclusions. Most articles that appear in journals have been revised once or twice prior to publication in response to critiques from the reviewers. Submitted articles that fail the review are, of course, rejected.[14]

The published scientific results considered by the IPCC have already

14. For readers interested in knowing more about peer review, please see a fuller discussion in my earlier book *Uncertain Science... Uncertain World* (Cambridge, UK: Cambridge University Press, 2003).

been peer-reviewed by the independent journals in which they were published, but that is not the end of peer review in the IPCC assessment reports. After a draft of an assessment report has been prepared, it is sent to a large pool of climate scientists not involved in its writing, but active in and knowledgeable about the fundamental science. They are asked to provide another layer of peer review to determine whether the draft assessment report is accurate, balanced, and free of distortion or exaggeration. Critiques can be formally expressed, and then forwarded to the assessment authors for a response. The authors are obliged to respond to each critique, either accepting and incorporating or rejecting and rebutting the essence of each commentary. More than thirty thousand written comments were submitted by more than six hundred individual expert reviewers of the Fourth Assessment Report's volume on the physical science of climate change.

The revised assessment report is next forwarded for review to the governments of the member countries represented in the United Nations. At this level the issues discussed are a blend of science, economics, and policy, but ultimately the language of the assessment reports must be approved by the governments. Considerable debate, accompanied by nuanced word-crafting, takes place, sometimes requiring an "agreement to disagree," but ultimately the text is approved and the assessment report becomes official and publicly available. For the Fourth Assessment Report in 2007 some 130 governments participated in this final stage of review.

The reason I have gone to great length to describe this review process is to make clear that, in the end, the IPCC report is a document that must, by any measure, be deemed conservative. The review process weeds out unbounded speculation, problematic science, and untested hypotheses. It carefully evaluates and states the uncertainties at every step of the way. In the end, what results is a lowest-common-denominator consensus of what the science is telling us. Moreover, the resulting reports are not policy prescriptive: that is, they do not tell governments what to do. They simply lay out various scenarios with attendant consequences: if you do X, you can expect Y; if you don't do W, you can expect Z.

The thorough and systematic quality control exercised by the IPCC contrasts strongly with the communications of the climate contras. These "skeptics" choose newspapers, radio, and television for their "scientific" pronouncements, or publications subsidized by vested interests that want to discredit what the real science is revealing. The contras have little interest in persuading the mainstream scientific community, and care little about peer review. The audiences they aim to persuade are state legislators and members of Congress, and other governing bodies around the world, where climate policy will ultimately be shaped.

THE VERDICT: "UNEQUIVOCAL"

So what did the IPCC's Fourth Assessment Report say about the evidence that Earth's climate is changing? Here is its bottom line:

> Warming of the climate system is unequivocal, as is now evident from observations of increases in global average air and ocean temperatures, widespread melting of snow and ice, and rising global average sea level.[15]

Ice everywhere is talking to us—not politically or emotionally or conventionally—but in a language that we must understand and heed. Ice is a sleeping giant that has been awakened, and if we fail to recognize what has been unleashed, it will be at our peril.

---

15. IPCC, "Summary for Policymakers," in *Climate Change 2007: The Physical Science Basis, Contribution of Working Group 1 to the Fourth Assessment Report of the Intergovernmental Panel on Climate Change*, ed. S. Solomon, D. Qin, M. Manning, Z. Chen, M. Marquis, K. B. Averyt, M. Tignor, and H. L. Miller (Cambridge, UK: Cambridge University Press, 2007).

The IPPC's use of the word *unequivocal* leaves little wiggle room. It means there can be no confusion about it. There can be no mistake about it. "Maybe, maybe not" is over. Significant climate change is happening. Seldom do we hear scientists make such an unambiguous pronouncement.

It is time to move on to other issues. Let us turn now to the causes of climate change. In the next two chapters we look first at the natural factors and then at the human factors that can cause Earth's climate to change on the time scale of the twentieth-century warming.

CHAPTER 5
NATURE AT WORK

The bright sun was extinguish'd, and the stars
Did wander darkling in the eternal space,
Rayless, and pathless, and the icy earth
Swung blind and blackening in the moonless air;
Morn came and went—and came, and brought no
day . . .

—LORD BYRON
"Darkness" (1816)

In early April of 1815 the city of Batavia, on the island of Java, began to hear sharp explosions that sounded much like distant artillery. But the city was not under attack, and no ship in distress was firing its cannons as a call for rescue. Batavia (now Jakarta) would later learn that it had been an aural witness to a violent eruption of the volcano Tambora, 1,200 miles to the east, along the Indonesian archipelago. For over a week Tambora erupted in a series of explosive events, creating sound waves heard even another 800 miles beyond Batavia. More than 70,000 people perished from the triple plague of a red-hot noxious gas and debris cloud rolling down the mountainside, a tsunami generated by the culminating explosion of April 11, and contamination of drinking water by the prolific ashfall. But if the devastating loss of life nearby was the immediate consequence, the eruption would continue to cause problems worldwide in the months and years to follow.

The explosions reduced the summit of the mountain from around 14,000 feet to 9,500 feet, blowing about 35 cubic miles of the volcanic edifice—an astounding volume of debris—into the atmosphere. In just a few months the atmospheric jet streams distributed this debris worldwide, blocking some of the sunshine from reaching and warming Earth's surface. The atmosphere is not quickly purged of such a burden, and the climatic effects were very apparent the following year. Temperatures throughout the Northern Hemisphere were depressed well below normal, and crop failures were common in Europe and North America. Connecticut experienced snow in early June, lakes in Maine froze over in mid-July, the mountains of Vermont were snow-covered in August.[1] Four killing frosts—one in June, one in July, and two in August—ensured that the New England harvest was meager. Around the world, the year 1816 became known as "the year without a summer."

Ash, dust, and chemical aerosols injected into the atmosphere during volcanic eruptions form a veil that blocks some of the incoming sunshine and prevents it from reaching and warming Earth's surface. The volcanic products have the effect of increasing Earth's albedo for a few years, reflecting more incoming solar energy back to space. Eventually, the ash and dust fall back to Earth, clearing the atmosphere and allowing the Sun's rays to once again warm the planet.

The eruption of Tambora was neither the first nor the last volcanic event to have a global impact on the atmosphere and climate. The diminished sunshine following an eruption in AD 536 on the island of

---

1. Henry Stommel and Elizabeth Stommel, "The Year Without a Summer," *Scientific American* 240 (1979): 176–86.

New Britain, just east of Papua New Guinea, led to this description of conditions in the Middle East:

> The Sun became dark and its darkness lasted for eighteen months. Each day it shone for about four hours, and still this light was only a feeble shadow. Everyone declared that the Sun would never recover its full light. The fruits did not ripen and the wine tasted like sour grapes.[2]

The dust from this eruption has been detected in well-dated ice cores in both Greenland and Antarctica. Tree-ring data from the European Alps, Scandinavia, and the Russian Arctic suggest that the cooling caused by this eruption may have been the most severe that the Northern Hemisphere has experienced in the last two millennia, even cooler than the effects from the 1815 eruption of Tambora.[3]

In August of 1883, another Indonesian volcano, Krakatoa, in one mighty explosion spewed more than six cubic miles of ash and dust into the atmosphere. The sound of the explosion was heard in Mauritius, an island located in the deep south of the Indian Ocean, some three thousand miles away, and the particles injected into the atmosphere soon led to spectacular red sunsets around the world. The colorful skies were captured in a series of sketches by the English painter William Ashcroft in 1884, and became known as the Chelsea sunsets. The official scientific report on the eruption of Krakatoa, published by Britain's Royal Society in 1888, featured the Ashcroft sketches as its frontispiece. For years following the eruption, brilliant sunsets and brutal winters were experienced around the world. The legendary harsh winter of

2. The author of these words is uncertain. They are likely from John of Ephesus, but have also been attributed to Michael of Syria. See the discussion in M. R. Rampino, S. Self, and R. B. Stothers, Annual Review of Earth and Planetary Sciences 16 (1988): 73–99.
3. L. B. Larson et al., "New Ice Core Evidence for a Volcanic Cause of the A.D. 536 Dust Veil," *Geophysical Research Letters* 35 (2008): article L04708.

1886–87 and the devastating blizzards of 1888—Krakatoa's unwelcome gifts to the struggling settlers and ranchers in the Great Plains—brought cattle-grazing on the open range of the United States to an end.

These examples make clear that great volumes of volcanic debris sent high into the atmosphere during an eruption, and soon thereafter distributed around the globe by the atmospheric circulation, can affect the global climate for several years. Volcanism is but one of the arrows in nature's quiver of climate-changing mechanisms that have played a role in Earth's climate in the aeons before humans populated the globe. Let us now look at some of the other processes that have altered Earth's climate prior to the appearance of humans.

WHEN THE CLIMATE is not changing, there is an equilibrium between the incoming energy absorbed by Earth's surface and the outgoing energy radiated back to space from the surface. All factors leading to climate change disrupt this balance between energy deposits and withdrawals from Earth's surface. Disruptions to the equilibrium include changes in the amount of sunshine arriving from the Sun, changes in the fraction of that energy that Earth reflects back to space, and changes in the atmosphere that cause it to capture some of Earth's heat instead of allowing it to radiate back to space unimpeded.

THE SUN DELIVERS

The amount of radiant energy Earth receives from the Sun changes over time, and not just because of variations in the amount that leaves the Sun. The periodic changes in the ellipticity of Earth's orbit around the Sun and in the tilt and precession of Earth's rotational axis—the Milankovitch cycles described in chapter 3—affect both the distance of Earth from the Sun and how Earth is oriented with respect to the

Sun. Being closer to or farther from the Sun will have obvious effects on Earth's temperature, and the changes in the tilt and orientation of Earth's rotational axis affect the strength of the seasonal variation in temperature.

But these Milankovitch cycles, with seemingly long periods of one hundred thousand, forty-one thousand, and twenty-three thousand years, are mere flutters on a very slow increase in solar luminosity that has been occurring since the beginning of our solar system four and a half billion years ago. This long-term increase in the Sun's radiance is a common evolutionary characteristic of millions of stars similar to the Sun. At the birth of the solar system the primeval Sun displayed only 70 percent of the luminosity it displays today. A dimmer Sun in that very early history of our solar system would imply that Earth was much colder in its early days, and ice much more common. Calculations of Earth's surface temperature with only 70 percent of today's solar radiation warming its surface inevitably translate into an ice-covered early Earth. And yet, geologists have identified widespread sedimentary rocks—rocks deposited in water—that are almost as old as Earth itself, suggesting that liquid H_2O was present early in Earth's history. This apparent contradiction was named the "faint young Sun paradox" by astronomers Carl Sagan and George Mullen.[4]

The paradox can be resolved with an atmosphere that also evolved in response to the slowly increasing solar radiation. Although nitrogen has probably always been the principal chemical component of Earth's atmosphere, oxygen has not. Oxygen in the atmosphere today is the principal waste product of photosynthesis, the process by which plants use sunlight to produce biomass—in other words, to grow. But photosynthesis did not become an important source of oxygen in the atmosphere until green plants evolved later in Earth's history. Initially there

4. C. Sagan and G. Mullen, "Earth and Mars: Evolution of Atmospheres and Surface Temperatures," *Science* 177 (1972): 52–56.

was very little oxygen in the atmosphere, because, absent photosynthesis, the only other process producing oxygen was a weak mechanism called photodissociation, in which radiative energy from the Sun broke the chemical bonds between the oxygen and hydrogen atoms in some of the water vapor molecules in the early atmosphere, thus freeing up a little oxygen. Even today, photodissociation yields much less oxygen than does photosynthesis, and under conditions of a dimmer Sun early in Earth's history, it would have been an even less efficient process.

With little oxygen available, carbon in the early atmosphere joined hydrogen to form methane, CH_4, in the atmosphere. Methane, however, is a potent heat-trapping gas, some twenty times stronger in its heat-trapping capability than its oxidized cousin carbon dioxide, CO_2. Earth's early atmosphere therefore acted as an extraordinarily effective blanket. As oxygen slowly became more abundant over geologic history, carbon dioxide gradually replaced methane, and the heat-trapping ability of the atmosphere slowly declined. When the Sun was weak, Earth's atmospheric heat-trapping blanket was strong, and as the Sun grew more radiant, the blanket grew weaker. The result is that Earth's average surface temperature has remained in the range of liquid H_2O throughout most of its history. The gradual oxidation of the atmosphere is now recognized as the resolution of the "faint young Sun paradox."

EARTH AS A GREENHOUSE

However much energy the Sun delivers to Earth, that energy can be diminished or enhanced by processes within our atmosphere. Explosive volcanism (as described earlier in this chapter) can block some of the solar radiation from reaching Earth's surface, and greenhouse gases in the atmosphere impede the escape of Earth's heat back to space.

How do the greenhouse gases trap heat trying to leave Earth? For a decent analogy, think of the microwave oven in your kitchen, in which microwaves (electromagnetic waves larger than infared but shorter than radio waves) are generated within the oven and absorbed by the food you want to heat. More specifically, the microwaves are absorbed by water molecules contained in the targeted morsels. Because the microwaves are absorbed within the food, the energy they carried as traveling waves is converted to another form of energy, heat, as required by the first law of thermodynamics.

Let's now scale the concept up to solar system dimensions to get a feel for Earth's natural (and long-standing) greenhouse effect. The dominant radiation the Sun generates is in the visible part of the electromagnetic spectrum, that band of wavelengths that our eyes have evolved to be sensitive to. These wavelengths include all the colors of the rainbow—red, orange, yellow, green, blue, and violet. The Sun sends off a little energy outside of the visible range—some shorter ultraviolet waves and some longer infrared waves—but most of the energy arrives in the visible wavelengths.[5]

Our atmosphere is essentially transparent to the visible wavelengths, so this energy from the Sun passes through the atmosphere unimpeded, to be absorbed at Earth's surface and warm it. But Earth cannot continually absorb energy and keep on heating up forever, at least not without serious consequences, such as melting. It must have a way of sending heat back into space to avoid continual warming. It accomplishes this balancing act by reradiating the energy received from the Sun back to space, but not in the visible wavelengths. The wavelengths that a body employs to radiate energy away depend on the temperature of the surface, with hotter bodies such as the Sun radiating shorter waves, and cooler bodies such as the planets radiating longer waves.

Energy comes to Earth as visible radiation from a very hot (~11,000°

---

5. If the sun were hotter and sending us more energy, it would shift its radiative peak to shorter wavelengths (we would then have evolved to see ultraviolet "colors"), and if it were cooler and sending us less energy, it would shift its radiation toward the infrared. In radiation physics, this relationship is called Wien's Law of Displacement.

Fahrenheit) Sun, but departs as invisible infrared radiation from a 60° Fahrenheit Earth. But now comes the hooker—the atmosphere, which is transparent to incoming visible radiation, is not fully transparent to the outgoing infrared waves. Several gases in our atmosphere, present in only tiny amounts, absorb infrared radiation and convert the radiant energy into heat. This is what we call the "greenhouse effect"—the process by which Earth takes in a little more heat than it sends back, and accordingly it must warm up a bit and radiate a little more, in order to restore the balance between incoming and outgoing energy.

The greenhouse effect is not simply some theoretical scientific construct—it is a very real observable and measurable phenomenon, and one we should be thankful for, because Earth would be much colder and inhospitable without it. The principal gases in the atmosphere that absorb infrared radiation are water vapor (H_2O), carbon dioxide (CO_2), and methane (CH_4); together they add up to less than 1 percent of the atmosphere. For every million units of atmospheric volume, only a few hundred parts are CO_2, and less than two parts are CH_4—but these minuscule amounts give a lot of "bang for the buck." One sometimes hears incredulity that such tiny concentrations can have any impact, let alone a major one. But these trace gases in our atmosphere raise Earth's surface temperature by more than sixty Fahrenheit degrees from what the surface temperature would be if Earth had no atmosphere. This natural greenhouse effect is what makes Earth the water planet, the blue planet, rather than just another of the many icy bodies of the solar system.

The absorption of infrared radiation by CO_2—an atmospheric process that would become so discussed in the second half of the twentieth century—was first measured by John Tyndall in 1859. It is no small historic irony that 1859 was the same year that petroleum was discovered in Pennsylvania by Edwin Drake. Little did anyone imagine that a century later the CO_2 from the combustion of petroleum would be warming the atmosphere by the mechanism first measured by Tyndall.

EARTH'S GEOLOGICAL THERMOSTAT

Earth has a "thermostat" that prevents the surface temperature from straying too widely. It does not, however, make adjustments daily, as in our homes. Rather, the adjustments take place over millions of years and are related to geologic processes that are temperature dependent. The conceptualization of this thermostat originated with Jim Walker, a very broad-based earth scientist at the University of Michigan. Walker synthesized perspectives from atmospheric science, oceanography, geology, and geochemistry to envision the way this geological thermostat works.[6]

Imagine an Earth that is a little too warm because of an atmosphere with an above-average concentration of the greenhouse gas CO_2. How does Earth turn down the thermostat? Some chemical reactions that decompose rock—a process that geologists call weathering—are more effective at higher temperatures, and so when Earth is warmer, the rivers that drain the continents carry a bigger load of dissolved chemicals to the sea.

One element that weathers from continental rocks is calcium, which, when delivered to the sea, combines with carbon dissolved in seawater to produce calcium carbonate, which ultimately is deposited on the seafloor as limestone. As carbon is removed from the seawater through limestone deposition, the sea pulls more CO_2 from the atmosphere, thus diminishing the greenhouse effect and cooling the planet. But as the surface cools, less weathering takes place, the supply of calcium to the sea slows, limestone deposition diminishes, and once again CO_2 builds up in the atmosphere to warm the planet..

Earth's temperature oscillates between warmer and cooler through fluctuations in the effectiveness of the natural atmospheric greenhouse,

6. J.C.G. Walker et al., "A Negative Feedback Mechanism for the Long-term Stabilization of Earth's Surface Temperature," *Journal of Geophysical Research* 86 (1981): 9776–82.

which in turn modulates the availability of calcium to form limestone. But the functioning of this thermostat is dependent on oceans full of water in which to absorb CO_2 from the atmosphere and deposit limestone on the ocean floor. Without water, Earth's thermostat would be broken.

VENUS—A PLANET WITHOUT A THERMOSTAT

Earth's closest neighbor in the solar system is Venus, the second rock from the Sun. Venus is similar to Earth in many ways: it is about the same size, its gravity field is about the same strength, its chemical composition parallels that of Earth, and it has an atmosphere. Its greater proximity to the Sun, at about only three quarters of Earth's distance from the Sun, would suggest a surface temperature warmer than Earth's, but still in the range that, if there were any H_2O present, it would be liquid water rather than ice or water vapor. So it was quite a surprise when planetary scientists discovered that the surface temperature of Venus was in excess of 860° Fahrenheit. This temperature is high enough to melt lead, and way too warm for water to exist at the surface of the planet.

What has happened on Venus? The clues emerge from the composition and mass of its atmosphere—a gaseous envelope around the planet nearly one hundred times more massive than Earth's atmosphere, and composed almost entirely of CO_2. In short, Venus has a thick greenhouse blanket that has trapped enough heat to raise the planet's surface temperature more than eight hundred Fahrenheit degrees higher than it would have without such an atmosphere. Comparatively, both Venus and Earth have a similar amount of carbon, but on Earth only a tiny fraction of the carbon is in the atmosphere. Most of Earth's carbon resides in deposits of coal, petroleum, natural gas, and limestone. In other words, Earth has stored most of its carbon underground, whereas carbon on Venus resides almost entirely in its atmosphere. What a difference that makes in the surface temperature!

Why is Venus unable to regulate its surface temperature in the way that Earth's geological thermostat has maintained a water-compatible temperature on Earth? The likely reason is that Venus is closer to the Sun, at a distance where it receives almost twice as much solar energy as Earth does. There, more intense evaporation led to a complete depletion of surface water, and without water there can be no biosphere to create coal, no oceans in which to deposit limestone—in short, Venus had no way to sequester carbon in solid form, and so carbon simply accumulated as gaseous carbon dioxide in the atmosphere, creating an intensely effective greenhouse blanket.

SHORTER-TERM FLUCTUATIONS IN CLIMATE

The slow evolution of the Sun over billions of years and the geological thermostat regulating temperature over millions of years cannot explain significant changes in climate over decades or a century. Those long-term natural processes change so slowly that effectively one century looks pretty much like the next. To explain the relatively fast contemporary warming in the twentieth and twenty-first centuries, we need to look for other causes. The variability of the energy radiated from the Sun has always been recognized as an important natural factor driving changes in Earth's climate.

Solar physicists and astronomers have learned from years of research that the Sun is a very active body, and that the amount of radiative energy leaving the Sun varies over many time scales—minute by minute, day by day, year by year, decade by decade. All of these rapid fluctuations ride along on the three long Milankovitch cycles, which create a slowly changing backdrop for the changes in solar output that occur on shorter and less reliably periodic time scales.

Short-term variations in solar output can impact Earth in a variety of ways. Occasional solar flares leaping upward a million miles above the Sun can be so intense that they disrupt radio transmissions on Earth for several days and even damage the electronics of orbiting communication satellites. These solar outbursts are a serious military concern, because they can blind battlefield surveillance from space and interrupt the remote control of the pilotless drones that are playing an increasingly important role in military operations.

The decadal variability of the solar output can be seen in the abundance of sunspots, dark irregular patches that appear on the face of the Sun. Sunspots have been observed astronomically for more than four hundred years.[7] Their numbers wax and wane with an apparent period of about eleven years—give or take a year. However, the strength of each cycle, as indicated by the number of sunspots at a cycle's peak, show upward or downward trends extending over centuries.

Generally speaking, the more sunspots there are, the more energy the Sun radiates. Conversely, when the Sun shows fewer spots, its radiative output is correspondingly diminished. The dark patches are actually colder than the brighter areas of the Sun that surround them, and so one might imagine that a darker Sun—one with more spots—would be radiating less energy than a brighter Sun. But the real meaning of more spots is that the Sun is churning more vigorously, and bringing more energy to its surface for radiation into space. When the Sun is "quiet," the spots are few, and outbound radiant energy is less.

In the period 1650–1715 there were very few spots on the face of

7. Counting sunspots day by day is yet another example of "Cinderella" science referred to in chapter 4. It is routine work, and yet the four-hundred-year archive provides valuable data for understanding solar variability.

the Sun, a period known as the Maunder Minimum, named for the nineteenth-century English astronomer Edward Maunder, who first pointed out the paucity of spots. On Earth it also coincided with a particularly cool interval within the broader climate downturn known as the Little Ice Age (mentioned in chapter 4), during which glaciers advanced in their valleys and the growing season grew shorter.

In the first half of the twentieth century, the peaks of the sunspot cycle grew for several cycles in a row. These decades of an increasingly active Sun probably contributed to a climb in the global average temperature of almost 0.9 Fahrenheit degree from 1910 to 1950. But in the last four decades of the twentieth century, the Sun has undergone a modest decline in radiative output. The minimum of the sunspot cycle in 2008 was the lowest in the past half century.

Since 1978, scientific satellites orbiting above our atmosphere have measured the incoming solar radiation in great detail—not only in the visible light of the electromagnetic spectrum, but also in the shorter ultraviolet and longer infrared wavelengths. These observations provide much more comprehensive information about the variability of solar radiation than does the simple counting of sunspots. The essential story the satellite radiometers tell, however, is the same as the sunspots: solar radiation has been declining in the latter decades of the twentieth century. Despite that, Earth's temperature has continued to climb over the same interval. Apparently, the Sun is sharing the stage with other factors that are affecting Earth's climate and causing it to warm.

VOLCANIC DIMMING IN THE TWENTIETH CENTURY

The eruptions of Tambora in 1815 and Krakatoa in 1883 were the signature volcanic events of the nineteenth century. The twentieth century was barely under way when the Santa María volcano in Guatemala erupted in 1902. This eruption blasted away much of the 12,000-foot summit of the mountain, sending some 1.3 cubic miles of volcanic ash high into the stratosphere and from there around the world. Although somewhat smaller than Tambora and Krakatoa, Santa María was estimated by volcanologists as probably one of the five biggest eruptions of the past few centuries. It was followed a decade later by the even bigger eruption of Novarupta, on Mount Katmai, in Alaska, which delivered more than four cubic miles of ash to the atmosphere.

The eruptions of Krakatoa, Santa María, and Novarupta in short order kept the atmosphere murky and the climate cooler for the better part of three decades. But after the 1912 eruption of Novarupta, there were no significant explosive volcanic events for a half century, which allowed the atmosphere to clear. The absence of volcanic dust also contributed to the climb in the global average temperature during the period 1910–50. But in the last half of the twentieth century, explosive volcanism returned. The eruptions of Agung in Indonesia in 1963, El Chichón in Mexico in 1982, and Pinatubo in the Philippines in 1991 kept the atmosphere dustier and the Sun dimmer than usual.

If these natural factors were the only ones at work in the last half of the twentieth century, a quieting Sun trying to penetrate a murkier atmosphere would have led to a slight cooling of Earth's surface. But in fact the temperature has continued to climb nearly one Fahrenheit degree since the mid–twentieth century, indicating that natural factors alone were not in control of Earth's climate. Indeed, for the first time in the history of Earth, other factors affecting climate—human factors—were growing in importance and beginning to overshadow the natural mechanisms.

Climatologists make a useful (albeit somewhat artificial) separation of the factors that cause changes in the climate, into natural and anthropogenic. Natural causes are those that are independent of human activity, whereas anthropogenic causes arise from human activity. It is safe to say that for most of Earth's history the causes of climate change were entirely natural, simply because there were no humans present on the planet. Our human predecessors, various species of the genus *Homo*, first appeared on Earth some three million years ago. As their numbers grew and their technology improved, their impact on Earth and the climate has become increasingly apparent.

In their 2007 Fourth Assessment Report, the IPCC scientists concluded with 90 percent certainty that the rise in temperature in the latter several decades of the twentieth century was attributable mainly to human activities. This ascendancy of the anthropogenic component of climate change, surpassing the natural drivers, was a subtle and unheralded tipping point in the history of our planet.

THE SECOND TRENCH OF DENIAL

In the previous chapter I note that there are people skeptical of the instrumental record of Earth's warming over the past century. These skeptics assert that the record misrepresents the true state of climatic affairs. They have argued that you can't believe the thermometers, or the scientists who deploy them and interpret the readings. This rejection of the instrumental record of rising temperatures was the first trench of denial the climate contrarians dug. They defended that trench tenaciously, but one by one they abandoned it, slowly retreating in the face of overwhelming evidence from human and natural thermometers that Earth was indeed warming.

But these climate contras soon set up camp in a second defensive trench—grudgingly accepting that Earth may be warming, but then arguing that humans have had nothing to do with it. If Earth is warming, they argue, then it must be due to the Sun, or to cyclical changes in Earth's climate associated with long-term variability in atmospheric and oceanic circulation. Even though the sunspot count and the direct measurements of solar radiation by satellites during the last half of the twentieth century both trend toward cooling, not warming, the skeptics have marshaled other arguments to support their belief that all climate change is solar in origin. Let's pause to examine some of these contrarian arguments.

Other planets are warming. One argument the contras put forward as "proof" that solar activity is driving climate change on Earth stems from changes observed on other bodies in the solar system. If several bodies are indicating a warming, then surely, according to this line of reasoning, the common cause must be the central element of the solar system—the Sun. The favorite example that the contras cite is the apparent warming of the (former!) planet Pluto by about 3.5 Fahrenheit degrees over the past two decades. The evidence of warming comes from an observed tripling of the atmospheric pressure of Pluto, which implies that some of the nitrogen at the surface of Pluto has evaporated and returned to the atmosphere. But if Pluto—the most distant large body in our solar system—has warmed by 3.5 Fahrenheit degrees because of increased radiant energy from the Sun, then planets closer to the Sun should have warmed even more. In particular, Earth—forty times closer to the Sun than Pluto—should have warmed more than 18 Fahrenheit degrees, an amount clearly far greater than Earth has experienced. If such a solar explanation for the warming of Pluto were true, there would be no ice left on Earth. A better explanation of the warming of Pluto can be found in seasonal effects in Pluto's 250-year orbital journey around the Sun, or possibly changes in Pluto's albedo that have led to less sunshine being reflected from its surface. Similarly, an apparent warming of Mars is almost surely due to fewer dust storms and a more transparent Martian atmosphere.

A cosmic ray connection. Another suggestion advanced by the climate contras relates to how the Sun might interact, via an intermediary mechanism, to change the amount of cloud cover over Earth, and thereby change Earth's albedo. This very complex scenario runs along these lines: Earth is perpetually being showered with cosmic rays from space—streams of charged particles that emanate from the Sun and other nearby stars. The particle stream from the Sun is called the solar wind. Most charged particles are deflected around Earth by our planet's magnetic field, or that of the Sun. But a few leak through the magnetic shield and are thought by some to promote clouds by serving as a "seed" around which water vapor will adhere and nucleate clouds. When the Sun is more active and generates a stronger solar wind, the magnetic shield contracts more tightly around Earth and becomes a better shield. Fewer particles leak into the atmosphere, and therefore there are fewer clouds. Thus a more radiant Sun would lead to lesser cloud cover over Earth, thereby allowing more sunshine to warm Earth's surface. Conversely, when the Sun is quieter, Earth's magnetic field relaxes a bit and allows more charged particles to enter the atmosphere and nucleate more clouds These reflect incoming sunlight back to space, which in turn will cool Earth's surface. The net result, were this complex scenario actually taking place, is that Earth would warm when the Sun is more active, and would cool when the Sun is quieter. Earth's temperature would rise and fall, tracking the ups and downs in solar activity. To the contras, this represents a possible mechanism that would reassert solar control of Earth's climate.

However, almost every element of this complex series of feedbacks is conjectural and unsubstantiated. Indeed, the nucleation effect of cosmic rays has not been demonstrated under realistic conditions in the laboratory, and cloud cover over Earth has not been observed to have a strong correlation with variations in the solar wind or cosmic rays in general. This mechanism gets high marks for imagination, but has not earned a passing grade in the real world of observations. It is an interesting idea, but there is no evidence to suggest that it actually works.

Natural cycles. The skeptics frequently assert that the current warming of Earth is the result of "natural cycles." They know that the geological record indicates swings of climate long before humans came to populate the Earth, and they suspect that maybe nature is again at work in the current warming episode. "Isn't today's climate change just one more example of these natural processes at work?" But the logic of this avenue of thinking is partially flawed, because the statement has an implicit but unfounded premise: the only factors influencing climate today are the same ones that have influenced climate in the geologic past.

This logical flaw can be easily seen with a simple analogy. Ask yourself: Did forest fires ever occur before there were people on Earth? The answer, of course, will be yes. Lightning strikes did start forest fires in the distant past. Then ask if that means that all forest fires today occur only because of lightning strikes. At that point the flaw in logic becomes clear: today, in addition to lightning, forest fires also result from arsonists, careless campers, and thoughtless smokers tossing cigarette butts from their cars. The takeaway lesson is that in addition to natural processes there is a new player in the forest fire arena today, the human population.

Credible climate scientists do not limit their inquiry into causes of climate change to those factors active in the prehuman past—they should and indeed do consider the possibility that over time, the causes of climate change may vary. Their task is to understand what has caused climate changes of the past, and what is causing contemporary climate change. The causes may or may not be the same, but scientists must evaluate the role of all possible causes, old and new, to decide which are the most important at a given time. And as the evidence does in fact indicate, human activities overtook natural factors in the twentieth century, to become the dominant force driving climate change today. Nature, long the conductor of the climate orchestra, has been displaced by the human population. In the next chapter we will see the great array of footprints we have placed on planet Earth.

CHAPTER 6
HUMAN FOOTPRINTS

I will bless you . . . and multiply your descendants into countless thousands and millions, like the stars above you in the sky, and like the sands along the seashore.

—GENESIS 22:17

IPCC scientists in their 2007 Assessment Report concluded that "most of the observed increase in globally averaged temperature since the mid-20th century is very likely (90% probability) due to the observed increase in anthropogenic greenhouse gas concentrations." In other words, according to the IPCC scientists, there are nine chances out of ten that we humans, through our burning of fossil fuels, have been the dominant factor in the warming of the last half century. Ninety percent certainty is an extraordinary statement of confidence in the conclusion—were you to go into a casino and be offered the opportunity to win at any game nine times out of ten, you would surely play with great confidence, and very likely leave with a bundle of cash.

But as certain as the scientists are about the role of humans in the recent climate change, the American public remains less persuaded. In a 2008 Gallup poll, only three out of five Americans believed that the climate was changing, let alone that humans had anything to do with it. The reasons why the American public has been slow to grasp the realities of climate change are many and complex, but certainly include the decades of disinformation and propaganda put out by the fossil fuel industry. Add to that eight years of the George W. Bush administration in Washington, which deliberately fostered additional doubts about climate change by exaggerating the scientific uncertainty and discouraging government climate scientists from speaking out about the causes and consequences of climate change. And there are a number of people simply distrustful of scientists because of the widespread scientific embrace of biological evolution that conflicts with their religious beliefs. So when scientists make pronouncements about Earth's changing climate, these same people dismiss the climate science because they don't trust scientists in general. They dismiss the message because of the messenger.

Certainly these industrial, governmental, and philosophical impediments have made it hard to persuade people that we humans have become big players in the climate system. However, other reasons also make it difficult for some people to recognize that the large human population has been driving Earth's climate away from the environmental background in which human society developed and thrived over the past ten thousand years.

Some find it hard to grasp the very concept that the global average temperature of Earth has changed over the past century. Most of us are unaccustomed to thinking at global spatial scales and intergenerational time scales. Whatever setting we are born into is imprinted upon us as normal and unchanging, even if it has experienced something different from the worldwide average and may be in the middle of rapid social, economic, and environmental change. We are not born with global vision or a sense of history. Sensing change over a time interval longer than the characteristic human lifetime requires a well-honed historical awareness and memory, attributes that no one is born with, and which therefore must be acquired.

A second reason why some people have difficulty recognizing a human role in climate change is that in our daily lives we tend to focus on contemporary local concerns. Some of this narrow focus stems directly from our evolutionary history. Only a thousand generations ago, our ancestors' chief occupation was the daily business of finding food in their immediate environment. Successful hunting of wild game and harvesting of natural fruits and grains were skills rewarded by natural selection. Humans did not have the need to know what the local climate would be like a century into the future, or whether there might be an intense drought developing halfway around the world due to an El Niño event in the Pacific Ocean. They were much more concerned with the necessities of the here and now, and had little time or inclination to ponder the abstract world.

Yet another reason why many people do not recognize their role in climate change is that their daily activities are separated from the subsequent effects those activities have on the climate, by both space and time. It is an abstraction to connect the simple act of increasing the setting of a thermostat in one's home, or driving alone to work each day, to the reality that these activities slowly but steadily increase the absorption of infrared radiation in the atmosphere and warm the planet.

But there is an even more fundamental reason that impedes recognition of the human role in climate change. In the face of hurricanes, tornadoes, tsunamis, earthquakes, and volcanic eruptions, all natural phenomena that can kill thousands of people very quickly, people feel very insignificant and powerless compared to the forces of nature. And indeed, as individuals, we do have very little power. But what people do not appreciate is that, collectively, the almost seven billion people on Earth today, with millions of big machines, are staggeringly powerful and becoming more so every year. It is the sum of activities of billions of individuals—a collective human force far greater than Earth has previously experienced—that is indeed changing Earth's climate.[1] Robert F. Kennedy understood this collective power when, in a social context, he said, "Few will have the greatness to bend history itself, but each of us can work to change a small portion of events and in the total of all those acts will be written the history of this generation."

In the remainder of this chapter, I guide you on a tour of our planet, and show you how completely we humans have taken control of Earth—its land, oceans, ecosystems, and, most certainly, its climate.

WHAT DO PEOPLE DO?

What on Earth have people been doing to push the climate out of equilibrium? One answer can be found in the way humans alter the land they live on, and in so doing change the planetary albedo—the amount of sunshine that is reflected back to space from Earth's surface. These changes to the land began long before the twentieth century.

The principal human activity that has directly led to changes in reflected sunlight is deforestation, whereby dark forest canopy is replaced by more open, lighter-colored, more reflective agricultural lands. Deforestation had a big head start over other human activities as a climate factor—with a beginning traceable to the human use of fire.

Fire was, and remains, nature's own agent of deforestation. Long before humans appeared on Earth, lightning strikes routinely set forests afire, and the flames burned until a lack of fuel or natural extinguishers—principally rainfall—eventually limited their spread. The arrival of humans did nothing to slow the burning; quite to the

1. Professor Iain Couzin, a mathematical biologist with appointments at both Princeton University and Oxford University, describes army ants in much the same way: "No matter how much you look at an individual army ant, you will never get a sense that when you put 1.5 million of them together, they form bridges and columns [with their own bodies]. You just cannot know that" (*New York Times*, November 13, 2007, D1).

contrary, early humans valued fire as a mechanism to drive and concentrate game, and a way to produce clear space where they could more easily become aware of nearby predators, and eventually to use it for agriculture. Once early humans discovered the advantages of fire for light, warmth, cooking, and protection, they worked hard to maintain and preserve fire rather than extinguish it. Fire was a friend, not foe, of the early hunter-gatherers.

Humans later "domesticated" fire—they learned to make fire with tools, to have fire when and where they wanted it. Fire, or the heat it generated, found new uses that eventually powered the industrial revolution. Controlled fires boiled water to produce steam for engines, and fires in confined spaces gave rise to the internal combustion engine—the burning of fuels within closed cylinders that made gas expand and push pistons to produce mechanical energy. Eventually, however, people began to think of fire not only as a friend, but also as a hazard, to the cities in which they lived and to the forest resource that provided construction material and fuel. Since about the middle of the nineteenth century, our attention has turned to extinguishing fire wherever it occurs unintentionally.

As Earth warmed following the end of the last ice age, the global population was much smaller than today. An estimated few million people were scattered over all the habitable continents, with a population density less than one person per square mile, only 10 percent of the density of present-day Alaska. But farmers these people were not. Everywhere, they remained dependent on hunting skills. And so formidable were these hunting skills that even the small human numbers were able to push the giant woolly mammoths, the mastodons, and the great Irish elk toward extinction.

The warming that followed the Last Glacial Maximum was episodic, but by ten thousand years ago the climate had become similar to what we, the present-day representatives of the human race, have known. Over the next ten millennia, almost to the present day, the climate remained remarkably stable at this new level, an equable condition that fostered fundamental changes in the way of life of humans. Climatic stability enabled the establishment of sustainable agriculture, which in turn provided sufficient food to allow population growth and urbanization, along with the specialized skills that develop in the urban setting. When a subset of the population can produce more food than they personally require, not everyone needs to be a hunter, or gatherer, or farmer.

USING THE LAND

Following the retreat of the continental glaciers, forests reoccupied the newly exposed terrain, eventually covering about one third of Earth's land area. With the establishment of agriculture, humans began to leave another big footprint on the landscape, cutting or burning the forests, plowing soil, and diverting water.

As successful agriculture supported a larger population, the many uses of timber accelerated deforestation. Not only did forests succumb to land clearing for agriculture, but increasingly timber also became an industrial commodity used in dwellings, urban construction, and even for roadbeds. There are still places in the world today where wheeled vehicles, motorized or otherwise, roll across the washboard-like surface of tree trunks laid side by side on the ground, mile after mile. Just a few years ago, in a visit to the temperate rain forest of Chile, I experienced such a roadbed, with its rhythmic staccato vibrations similar to those that accompany travel on a corrugated gravel road.

Another use of timber led to equally dramatic deforestation. The recognition of the simple fact that wood floated on water stimulated the large-scale building of sailing ships for exploration, colonization, trade, piracy, and political and military advantage. The Phoenicians, Romans, and Vikings sailed long distances in substantial wooden ships. The

European powers of the Middle Ages became the first practitioners of globalization, sending vessels around the world to disseminate Christianity and accumulate wealth. The 1571 eastern Mediterranean battle of Lepanto, between the European Holy League and the Ottoman Turks, and the failed invasion of England by the Spanish Armada near the end of the sixteenth century both involved hundreds of naval vessels constructed of prime timber, each requiring thousands of mature trees. The forests of Europe no longer seemed infinite.

When the Europeans arrived in North America, about 70 percent of the land east of the Mississippi River was forested. By the end of the nineteenth century that had been reduced to around 25 percent. Much of the landscape was literally stripped naked. Wood was used for nearly every endeavor in the growing nation—for the barges and channels and locks of the inland canal system; for the ties, trestles, and rolling stock of the national railways; for the fences that demarked property; for the telegraph and telephone poles that enabled early telecommunication; and eventually for paper. Photos of the state capital in Vermont toward the end of the nineteenth century show the extent that the hills surrounding Montpelier had been denuded. Such scenes were widespread across North America—almost the entire forest cover of Michigan succumbed to logging, and to fires that often followed close on the heels of careless lumbering practices.

Today, deforestation remains very active in many parts of the world. The tropics are being particularly hard hit, with large areas of Brazil, Indonesia, and Madagascar being subjected to relentless clear-cutting. Half of the world's tropical and temperate rain forests are now gone, and the current rate of deforestation exceeds one acre per second. That is equivalent to cutting down an area the size of the state of Mississippi each year. But in places such as the eastern states of America, where deforestation was rampant in the nineteenth century, the forests are returning, as other materials have replaced wood in much of the modern economy of the region. The recovery in the eastern states, however, is far from complete—today's second-growth forests cover only 70 percent of the pre-colonial distribution.

How do changes in the forest cover affect climate? Deforestation generally changes the color of Earth's surface from dark green to lighter brown, thus causing more sunshine to be reflected back to space rather than warming the planet's surface. Countering this slight cooling, however, is the much more significant effect of the cutting and burning of trees itself. In the natural state of affairs, living trees pull the greenhouse gas carbon dioxide (CO_2) from the atmosphere in the process of photosynthesis, and dead trees decay, liberating CO_2 and returning it to the atmosphere—an atmospheric equilibrium established by pulling out and pumping back equal amounts of CO_2. But rapid and large-scale deforestation upsets that equilibrium—the loss of trees decreases photosynthesis, leaving more CO_2 in the atmosphere. And when deforestation occurs by burning, it returns CO_2 to the atmosphere far faster than new trees can grow and remove it. In sum, deforestation leads to warming of the atmosphere.

HUMAN NUMBERS GROW

The past ten millennia have generally been good times for us humans, and we have multiplied at a breathtaking pace. At 6.8 billion and growing, the human population today is more than a thousand times bigger than it was at the end of the last ice age, some 10,000 years ago. But the growth of population has not been steady over that time—it has accelerated dramatically in recent centuries.

Multiplying a number by one thousand is almost the same as doubling that number ten times. The concept of ten doublings is a good approximation of the growth of Earth's human population from the last gasps of the ice age, when the population was around four million people, up to almost the present day. The growth began slowly—the first, second, and third doublings together required more than six thousand years, an interval of time that began when humans first began to congregate in villages and ended not long after the construction of the Great Pyramids of Egypt. The fourth doubling required a thousand years, and the fifth, only five hundred. The sixth doubling began when Rome ruled the West and the Han Dynasty the East, and ended as Europe entered the Dark Age. The seventh doubling took place in the seven hundred years between 900 and 1600, slowed by the Black Plague, which killed a quarter of the global population in the fourteenth century. The seventh doubling ended just as European explorers were circumnavigating the globe and claiming colonial territory in the New World. The eighth doubling, occurring in the two hundred years between 1600 and 1800, encompassed the creation of the United States of America, and carried the global population to the landmark statistic of one billion human inhabitants.

An extraordinary change in technology also occurred during the eighth doubling: the discovery of how to access the fossil energy contained in coal. No longer would humans rely solely on wood for heat or flowing water for industrial power. Spurred on by the abundant energy in coal, the world population underwent its ninth doubling in only 130 years, to reach two billion by 1930, in spite of the Napoleonic Wars, World War I, and a virulent flu pandemic. The tenth doubling occurred between 1930 and 1975, overcoming the effects of World War II and three subsequent Asian wars. Those ten doublings took Earth's population from four million around ten thousand years ago to four billion in 1975, and the doubling interval shrank from twenty or thirty centuries to fewer than five decades. The eleventh doubling, now under way, from four to eight billion, will be achieved around 2025.

As I write in early 2009, the global population is at 6.8 billion. Were a person to be born each second, and if no one ever died, it would take more than 215 years to populate Earth with 6.8 billion people. The current rate of population growth is more than a million people each week, the result of more than 4 births, offset by fewer than 2 deaths, each second. At that rate, Earth's population grows by the addition of a Philadelphia or a Phoenix each week, a Rio de Janeiro each month, and an Egypt each year.

A doubling of Earth's population, a process that once required a few thousand years, today takes place in less than fifty. The human footprint on the planet is increasingly apparent simply due to the sheer number of people on Earth today. One cannot fully understand the changes in the global environment under way outside the context of the dramatic population growth of the last few centuries.

PEOPLE AND MACHINES

Since the end of the last ice age, humans have grown not only in numbers, but also in technological skill and resource consumption. In only a thousand generations, they have moved from human power to horsepower, at first literally and later with machines that amplified the strength of humans and their domesticated beasts of burden. These machines have enabled us to travel far faster than we or horses can run, carry far more than the capacity of backpacks or saddlebags, dig far deeper in the soil than shovels, hoes, or plows can reach, and kill far more people faster than clubs, spears, or arrows could ever accomplish.

For much of the industrial revolution, the rate at which humans use energy was measured in horsepower—a throwback to one of the animals that humans domesticated for agriculture and transportation. That unit of energy expenditure remains in common use in the automobile industry, where the power of engines is still rated in horsepower. James Watt, the developer of one of the first commercial steam engines, wanted a way to compare the work his engine could accomplish to the power output of the more familiar workhorse. Watt estimated the lifting that one horse could accomplish in bringing coal out of a mine. He determined that a horse could, using ropes and pulleys, lift a ton of coal up a mineshaft fifteen feet each minute, which, when expressed in the more common terms for the rate of energy use, is about 750 watts.[2] This is about the power required for a small microwave oven or space heater. The kilowatt-hour is the common unit of electricity consumption, which translates into using electricity at a rate of a thousand watts for one hour, or a little more than one horsepower for an hour. In my home, my family and I consume around twenty-four kilowatt-hours of electrical energy each day, which is the equivalent of having a horse working around the clock.

Of course, we use much more energy in our daily lives than just electrical energy. There is natural gas used to heat my home, gasoline used in the car I drive to and from work, and energy used in my workplace. Energy is also used for manufacturing, bulk transport of goods, agriculture, and much more. Effectively we all have many more horses working for us. Worldwide, the per capita rate of energy consumption is about 2,600 watts—that is, about 3.5 horsepower for every man, woman, and child on the planet, or the energy equivalent of a global population of workhorses numbering almost 25 billion. And the rate of energy consumption is hardly stable—to the contrary, it increased sixteenfold during the twentieth century alone.

Surely many people in developing countries would welcome the news that they have three and a half horses working for them. But of course the global average rate of energy consumption is deceiving—many people have no horses at all working for them, and some others have a stable full. In the United States, the 300 million residents, about 4 percent of the world's population, account for 20 percent of the global energy expenditure. That amounts to more than fifteen horses working for every single American.

2. The term *watt*, honoring James Watt, describes the rate of energy use. One watt is equal to one *joule* (a unit of energy in the metric system) per second.

Richard Alley, a well-known climate scientist at Penn State University, has carried the horse analogy further. He points out that the carbon emitted from the combustion of fossil fuels is in the form of the greenhouse gas carbon dioxide (CO_2). This gas is colorless, odorless, and tasteless, and so its presence in the atmosphere is not easily detected with our human sensory organs. But Alley asks us to imagine how different our attitude toward this important source of global warming would be if the carbon were emitted not as an invisible gas but rather as horse manure that accumulated ankle deep over the entire land surface. That would certainly get our attention in ways that CO_2 in the atmosphere does not.

PLOWING AND BUILDING

Deforestation was only the beginning of human interactions with the natural Earth. Once people cleared the land for agriculture and towns, they put sticks, bones, spades, plows—and later tractors, bulldozers, steam shovels, and massive excavators—to work. As earth-movers, humans showed what they could do, and they could do a lot,[3] century after century.

The seeds of agriculture were first sown some nine thousand years ago, as villages became established and nomadic life gave way to a more rooted, sedentary social structure. About 2.5 acres of crop and pastureland were required to feed a person for a year then, and it is not much less even today. Every year, the loss of topsoil associated with tilling the land, at least until the adoption of soil conservation measures in the mid–twentieth century, amounted to about ten tons for each person, or

3. The short article "On the History of Humans as Geomorphic Agents," by Roger LeB. Hooke (*Geology* 28, no. 9 [September 2000]: 843–46), provides a very readable summary of the human shaping of Earth's solid surface.

about the volume of ten human graves for each person fed by agriculture. What has changed dramatically, of course, is the number of people to feed. With the global population nearing seven billion people, we lose on average about three inches of soil to erosion every century over all the farm and pasture land of the world, an area close to 40 percent of Earth's ice-free land surface.

As people developed quarrying and mining, both for raw materials and energy, they dislodged more and more earth. With urbanization also came the need for water for the growing population, thus leading to the excavation of canals and the construction of aqueducts. Political and economic control required road and wall building—the Romans paved nearly two hundred thousand miles of roads and highways, and built Hadrian's Wall seventy-five miles across the north of England as a defense against the unwilling-to-be-governed Scots. The Chinese built the Great Wall—actually a series of walls—stretching for some four thousand miles across northern China to defend against Mongol raiders. Great monuments, such as the pyramids of Egypt, and less grandiose but widespread burial mounds constituted massive construction projects.

In the modern world, the scale of our human assault on the landscape is no less profound. Coal mining, always a hazardous operation underground, surfaced with the discovery of widespread coal deposits with just a thin veneer of soil covering them. Surface strip-mining increased the volume of coal, rock, and soil moved by a factor of ten or more compared with underground operations. Today in the Powder River Basin of Wyoming, gigantic machines claw into the thick coal seams, delivering load after load of coal to waiting railway hopper cars. Mile-long railway trains leave the mines every twenty minutes. Over a year the trains could form a belt that completely encircles Earth. These trains snake across the prairies of Nebraska in an unending stream, slowly diverging to other tracks that fan the delivery of coal to electrical power plants in the eastern and southern states. And the pits of Wyoming—the residual scars of strip mining—grow larger.

In the east, coal mining continues in Pennsylvania and West Virginia. The long-ago geological collision of Africa with North America folded the coal seams, along with the other sedimentary layers, into the beautiful valley-and-ridge topography of the Appalachians. Erosion along the crests of the ridges has brought the coal seams closer to the surface, but not quite unburdened them to full exposure. But that is no insurmountable problem to coal mining today—just move in with dynamite and earthmovers, scrape off the mountaintops until the coal is exposed, and strip the coal away.[4] The overburden, as geologists call the rock that sits between the coal and the surface, is unceremoniously dumped into the adjacent valleys, where it destroys forests on the mountain slopes and causes flooding in the streams occupying the valleys below. The very name of this process—mountaintop removal—expresses both the power and hubris of this human endeavor.

Giant earth excavator[5]

4. To read more about mountaintop removal mining, see "Mining the Mountains," by John McQuaid, in the January 2009 issue of *Smithsonian*.
5. The illustration is after a photo of this excavator by Dermot McArdle, Fleet General Manager, KMC Mining.

ALL OF THESE changes brought to the landscape by industrious people are best understood and more fully appreciated only when they are compared with natural processes that move earth around. When compared to nature's ability to move sediment and rock, is the human impact trivial or enormous? This question fascinated Bruce Wilkinson, one of my geology colleagues at the University of Michigan for many years. Bruce is not a dapper tweed-attired professor—he is a gritty field geologist, always in Levi's; with his big voice, he is never afraid to speak truth to power or call attention to a naked emperor. Not surprisingly, Bruce approached the question of whether human earth-moving was significant with geological logic. He reasoned that the long-term record of sediment erosion and transport can be found in the sediment deposits that have accumulated on the ocean floor, and in the sedimentary rocks of past eras on the continents. He calls this the "deep-time perspective."[6] It involves the careful estimation of sediment volumes: in the deltas of all the rivers of the world, on the continental shelves, on the deep ocean floor, and in the ancient sedimentary rocks now stranded on the continents.

Wilkinson's calculations showed that over the past five hundred million years, natural processes of erosion have on average lowered Earth's land surface by several tens of feet each million years. When he next calculated the present-day rate of erosion, the result was startling—humans are moving earth today at ten times the rate that nature eroded the planetary surface over the past five hundred million years. Perhaps more alarming is the erosion rate in the places where the erosion is actually occurring. On land used for agriculture, soil loss is progressing at a rate almost thirty times greater than the long-term worldwide average of natural erosion.

6. Bruce H. Wilkinson, "Humans as Geologic Agents: A Deep-Time Perspective," *Geology* 33, no. 3 (March 2005): 161–64.

Not only is soil erosion much more rapid than in the geological past, but the rate of soil loss also far exceeds the rate at which new soil is produced.[7] In the same way that our consumption of petroleum far exceeds the pace at which nature makes it, and our withdrawal of groundwater far exceeds the rate at which nature recharges aquifers, the human practices that lead to the loss of agricultural soil are effectively "mining" the soil, using up a resource of finite extent. As humorist Will Rogers once noted, "They're making more people every day, but they ain't makin' any more dirt." David Montgomery of the University of Washington estimates that almost a third of the soil capable of supporting farming worldwide has been lost to erosion since the dawn of agriculture, with much of it occurring in the past half century.[8]

Once the land surface is plowed for agriculture, or opened to livestock grazing, wind has easier access to dust to blow around. The blowing dust is dropped eventually, and some falls into lakes and the ocean. The amount of dust accumulating in lakes of the western United States has increased by 500 percent during the past two centuries, an increase attributable to the expansion of livestock grazing following the settlement of the American West.[9]

Blowing dust travels the world. Satellite photos show huge clouds of dust billowing out of the Sahara Desert in northern Africa, in giant plumes that spread westward over the Atlantic Ocean. From China, similar clouds head eastward across the Pacific, but unlike the clouds from the uninhabited Sahara, the clouds emanating in China are not just dust—they include industrial pollution that the winds carry all the way to the western coast of North America. The atmosphere, by distributing

7. B. H. Wilkinson and B. McElroy, "The Impact of Humans on Continental Erosion and Sedimentation," *Geological Society of America Bulletin* 119 (2007):140–56.
8. D. R. Montgomery, "Is Agriculture Eroding Civilization's Foundation?" *GSA Today* 17, no. 10 (2007): 4–9.
9. J. C. Neff et al., "Increasing Eolian Dust Deposition in the Western United States Linked to Human Activity," *Nature Geoscience* 1 (March 2008): 189–95.

industrial waste and unintentional erosion from agriculture, is an effective agent of globalization—the globalization of pollution.

Blowing dust and soot from diesel engines, cooking fires in rural undeveloped areas, and the burning of grasslands and forests are also having their effects on climate, both regionally and globally. In the same way that ash from large volcanic eruptions blocks sunshine from reaching Earth's surface by making the atmosphere less transparent, dust and soot also dim the Sun, at least as it is seen from Earth. But the dark soot particles in the atmosphere also absorb some of the sunlight reflected back to space from Earth's surface, thereby trapping energy in the visible wavelengths of the solar spectrum just as greenhouse gases absorb some of the infrared wavelengths. The dust and soot also lead to accelerated melting of snow and ice around the world, by darkening the white surface ever so slightly. The darkening causes less sunlight to be reflected back to space and more solar heat to be absorbed by the dust and soot, thereby further increasing the melting of the snow and ice.

FLOWING WATER

It is not just the land that people have changed—they have had equally dramatic effects on the water. The development of agriculture and urbanization could not have proceeded without the parallel development of water resources. The roots of hydraulic engineering date back almost six thousand years, and these special skills developed independently in many locations. In ancient Persia, large underground conduits called qanats carried water from the highlands to the arid plains. People built levees to stabilize river channels, and canals to carry water to fields for irrigation along the banks of the Nile in Egypt, the Indus in Pakistan, and the Yellow River in China. In the Fertile Crescent of Mesopotamia, water management along the Tigris and Euphrates reached high levels of sophistication thousands of years ago.

Following the retreat of the last continental ice sheets, many areas of North America and Europe were dotted with small lakes, marshes, and swamps. And as sea level rose following deglaciation, the estuaries of rivers extended farther inland, creating additional wetlands. Agriculture demands, land development pressures, and public health concerns led to the draining of wetlands. The District of Columbia, seat of the United States government, was originally a malarial swamp, as was much of southern Florida. The eradication of malaria in the United States was a singular public health achievement resulting from the draining of wetlands.

Today, half of the wetlands that existed around the world only ten thousand years ago are gone. Although there have been some genuine benefits associated with wetland drainage, there have also been losses. When wetland drainage began in earnest, little was known of the many services wetlands provided—their importance in the ecology of wildlife and the role they played in water purification and as a buffer against storm surges in coastal areas. And the assault on wetlands is not over—in the United States, wetland loss continues at a rate of one hundred thousand acres every year.

We have left our imprint on the lakes and rivers of the world as well. The Aral Sea, a once huge inland body of water situated along the border between Kazakhstan and Uzbekistan, in central Asia, was only half a century ago high on the list of the world's largest lakes, surpassed in area only by the upper Great Lakes of North America and Lake Victoria in Africa. Today the Aral Sea has almost disappeared, reduced to a scant 10 percent of its former area.[10] The shrinking has not been due to long-term climatic cycles that sometimes lead to lake fluctuations elsewhere. No, the Aral Sea has been the victim of water diversion from the two principal rivers that feed it, water diverted to irrigate cotton planted in the desert of Uzbekistan. Diversion canals began taking water away

10. Philip Micklin, "The Aral Sea Disaster," *Annual Review of Earth and Planetary Sciences* 35, no. 4 (2007): 47–72.

from the Aral Sea following World War II, and by 1960 the lake level began to fall—almost a foot each year in the 1960s, but tripling by the end of the century as withdrawals grew.

As the lake level fell and the water volume diminished, the remaining water became more saline, in much the same way that Great Salt Lake of Utah became saltier as it shrank from its glacial meltwater maximal extent. Today what remains of the Aral Sea has a salt concentration ten times its pre-diversion salinity, and three times the salinity of ocean water. The fishing industry of the Aral Sea, formerly producing one sixth of the fish of the entire Soviet Union and employing tens of thousands of people, disappeared with the water. Abandoned fishing boats now sit motionless on a sea of sand.[11]

RIVERS HAVE FROM the earliest days of human settlement been favored places for villages. They provided domestic and agricultural water, avenues of transportation, and power for industrialization. The human tendency to "control and improve" nature led to the construction of dams on many rivers. Today, some fifty thousand large dams and many smaller ones have altered the natural flows of rivers virtually everywhere. No major river in the United States has escaped damming somewhere along its course; not a single one flows unimpeded from headwaters to the sea.

In the spring of 1953, when I was in high school in Omaha, a major flood began to build farther up the Missouri River as heavy snow melt and ice dams raised this great river to flood stage. As the crest approached Omaha, it became apparent that it might top the levees and floodwall that protected the lower levels of the city, including the airport. The call went out for volunteer manpower to sandbag the levees. Classes be damned—this was an opportunity for students to get involved in some excitement,

11. The worst of the Aral Sea decline may be in the past. The diversions of water have recently been partially reversed, and the lake level shows signs of stabilizing.

and so many of my friends and I were soon patrolling and reinforcing the levees as the water crept upward. When the crest was only hours away, it became obvious that it might top the floodwall, so emergency construction began to place flashboards atop the floodwall—and they proved essential. The river crested two feet above the floodwall, but was held back by the flashboarding. Omaha was saved from a major flood, ironically the last flood ever to roll down the upper Missouri River Valley. Soon after, six large dams were constructed upriver, in the Dakotas and Montana, that have ended the Missouri River's free-flowing days.

Later in my professional career as a geologist, I rafted the length of the Grand Canyon of Arizona, a chasm cut by millions of years of erosion by a turbulent Colorado River. To every serious geologist, the Grand Canyon is an obligatory pilgrimage, a geological Mecca. When first explored by John Wesley Powell, an early director of the U.S. Geological Survey, in the 1860s, the Colorado River was free-flowing, with treacherous rapids and large seasonal changes in volume. Today the river that carved the greatest canyon in the world is just a controlled-flow connection between the huge Glen Canyon and Hoover dams, two of the five large dams on the Colorado. In the Grand Canyon, the Colorado River predictably rises and falls on a daily basis, as water is released from the Glen Canyon Dam each day. Ironically, even as the flow has steadied, some of the rapids have become even fiercer. Within the canyon, where tributaries join the Colorado, they continue to deliver debris that no longer is moved downstream by an annual spring flood on the Colorado. The result is a growing fan of boulders spilling into a narrowing river channel—a sure recipe for enhancement of rapids.

A few years ago I traveled on a small boat down the Douro River in Portugal.[12] The Douro arises in northern Spain (where it is called

---

12. For the fascinating story of this trip, read John Pollack's *Cork Boat* (New York: Anchor Books, 2004).

the Duero) before crossing into Portugal on its way to the Atlantic. The stories of the early navigators of the Douro tell of cataracts, turbulence, and danger. Much to my surprise, the Douro today is little more than five narrow flat lakes behind five large dams. Whatever current exists today is due to controlled flow through the locks and spillways of the dams.

Dam construction is hardly a thing of the past. The Itaipu Dam, on the Paraná River of South America, became operational in 1984. When it opened it was the largest in the world in terms of electrical generating capacity. Itaipu's electricity output has since been overtaken by the massive Three Gorges Dam on the Yangtze River in China, already impounding water and scheduled for full operations in 2012. Three Gorges is only the latest of major dam construction projects on all the principal rivers of southern Asia, and China has plans for another dozen in the upper reaches of the Yangtze.

Dams and diversions so diminish the flow of major rivers that some rivers barely reach the sea. The apportionment of Colorado River water between the United States and Mexico (the river's mouth is in Mexican territory) was decided on the basis of measured river volumes of the early 1920s. Unappreciated at the time was the fact that the river volumes then were at historic highs, not to be seen again in the rest of the twentieth century. Today, after the allotted withdrawals from the Colorado River in the United States, there is little water left for Mexico, and in some years not a drop of water flows out of the mouth of the Colorado into the Gulf of California.

The situation is not very different on the Ganges or the Nile, where the flows at the river mouths have been reduced to just a trickle. The fertile Nile Delta, long the principal source of food for much of Egypt, is being slowly reclaimed by the Mediterranean Sea because so little sediment is carried by the weakened Nile to replenish the soil of the delta. Farther upstream, where the Nile now flows slowly, and in the giant reservoir behind the great dam at Aswan, where it hardly flows at all, quiet backwaters have been created that have allowed schistosomiasis, a debilitating parasitic infection, to flourish in places it never appeared before. Schistosomiasis is second only to malaria in terms of tropical diseases that afflict humans.

The sediment load is not the only burden moved along by rivers. They also carry heavy chemical loads, picked up from industrial pollution, inadequate sewage treatment systems, increased runoff from the impervious surfaces of urban areas, and the fertilizers and pesticides used in agriculture. More manufactured nitrogen fertilizer is spread over agricultural lands than is provided by the entire natural ecosystem. Although the sediment load is slowed in its passage to the sea by the presence of dams, the chemical load carried by rivers is virtually unaffected by those barriers. Effectively, the chemical waste streams of entire continents are flushed into the sea.

This flux of chemicals to the sea is increasingly producing "dead zones" in the oceans, regions of the ocean floor devoid of all but microbial life forms. The delivery of fertilizers to the sea promotes the growth of algae in the surface waters, but when these organisms die, they fall to the seafloor, where they fuel microbial respiration. The dissolved oxygen in the bottom water is depleted by the microbes, and is therefore unavailable for other bottom-dwelling marine creatures, principally fish and mollusks, that need oxygen.

Dead zones are appearing all over the globe—in Chesapeake Bay; the Gulf of Mexico; the Adriatic, Baltic, and Black seas of Europe; and along the coast of China. The number of dead zones, now in excess of four hundred, has roughly doubled every decade since the 1960s. In aggregate they now cover an area approaching one hundred thousand square miles,[13] about the size of the state of Michigan.

---

13. Robert J. Diaz and Rutger Rosenberg, "Spreading Dead Zones and Consequences for Marine Ecosystems," *Science* 321 (August 15, 2008): 926–29.

WATER UNDERGROUND

Large numbers of people in the United States drink water that is pumped from underground. And not all of these people live in rural areas—many cities also draw some of their municipal water from wells. Almost 40 percent of the public water supply in the United States is pumped from subsurface aquifers, and nearly all residential water in rural areas is drawn from a well. But domestic use of groundwater represents the lesser call on underground water—twice as much is pumped for agricultural irrigation in the areas of the United States where precipitation is inadequate, at least for the crops selected for cultivation. The regions most dependent on groundwater for agriculture are in California and the Southwest (where there is a great use of the surface water as well) and in the vast Great Plains of Montana, the Dakotas, Nebraska, Kansas, Oklahoma, Texas, New Mexico, and Colorado.

On the Great Plains, the 100° meridian of west longitude is the dividing line for irrigation—to the east there is usually sufficient winter snow and summer rain to keep the soil moist enough to support agriculture naturally. West of longitude 100° the crops need some help from irrigation. Fortunately, thick layers of sand and gravel, representing millions of years of waste and wash from the eroding Rocky Mountains, lie beneath the surface of the Great Plains. The never-ending competition between tectonic uplift of the mountains and the erosive power of rain, snow, and ice produces vast amounts of debris, carried eastward by streams and rivers meandering across the plains. And within the buried deposits of sand and gravel, water fills the pore spaces between the grains, water that had not seen the light of day for many thousands of years—until mechanized agriculture came to the Great Plains in the twentieth century.

The prairie pioneers overturned the sod and dug wells to withdraw water from the saturated aquifers beneath them. The first mechanical boost to pumps came from windmills, which took advantage of the wind that came "sweeping down the plain." Rural electrification provided steady energy independent of the vagaries of the wind, and soon pumps of higher capacity were pulling more water from the aquifers below.[14] The best known of these buried aquifers is called the Ogallala, the name of a small town in western Nebraska. More than a quarter of the irrigated land in the United States sits atop this aquifer.

Withdrawals of water from the Ogallala aquifer have been at a far faster rate than nature has been able to recharge it—leading to a drop in the subsurface water table. In areas of intense irrigation, such as in southwestern Kansas and west Texas, the water table has been dropping several feet each year, requiring wells to be deepened just to reach the water. In some places the water has been exhausted. Effectively, the water that had been in place since the end of the last ice age has been "mined out." With the draining of this aquifer at a rate equal to some fifteen Colorado Rivers each year, natural recharge cannot keep pace—the groundwater in reality is a non-renewable resource.

WILL WE TAKE IT ALL?

Yet another measure of our footprint on Earth is the fraction of the total plant growth on the continents that we have appropriated to serve our needs for food, clothing, shelter, and other necessities and amenities of life. Before humans became a presence on Earth, the fraction appropriated was zero. But today, with more than 6.8 billion people occupying the continents, the fraction is certainly not zero. This measure of the human dominance of the land-based ecosystems is variously estimated

14. Ironically, today wind turbines are being installed on the plains to generate the electricity for irrigation pumps—a return of wind power to a setting where it had been largely neglected for half a century.

at between 20 and 40 percent of net primary production,[15] an astounding number in spite of the uncertainty.

The appropriation of resources for human use is not restricted to the land. In the oceans, an estimated 90 percent of the large predator fish present in the oceans a half century ago are gone. Although there are ample dry statistics to make this point, none is quite so visual as an archive of tourist photos from Florida showing the size of the catch worth bragging about, progressing through the twentieth century. Decades ago the catch worth posing for was as big as the fisherman, but through the years photos reveal smaller and smaller fish, with today's trophy catch seldom exceeding three feet.

Three quarters of the marine fisheries around the world are now fished to capacity or overfished. Thirty percent of the fisheries have collapsed, a term defined to indicate a depletion greater than 90 percent of the recorded maximum abundance.[16] Cumulative catches of the global fishing fleet reached a maximum in 1994, and since have declined by 13 percent, despite large increases in the fleets and their range of operations. A well-recognized indicator of marine resource depletion is the fact that the fish catch is declining despite many efforts to increase it.

Human depletion of other resources can be recognized in the same way. Petroleum production in the United States reached a peak in the mid-1970s and has declined ever since, in spite of the fact that the petroleum industry has intensified exploration and has been motivated by substantial increases in the price of oil and natural gas. Worldwide, the production of petroleum will likely reach a peak in the next decade or two.

15. Haberl, H., K.-H. Erb, and F. Krausmann, "Global Human Appropriation of Net Primary Production," *The Encyclopedia of Earth*, 2008, accessed at http://www.eoearth.org/article/Global_human_appropriation_of_net_primary_production_(HANPP).

16. B. Worm et al., "Impacts of Biodiversity Loss on Ocean Ecosystem Services," *Science* 314 (2006): 787–90.

LEAVE ONLY FOOTPRINTS

As tourists, we have been urged to "take only pictures, leave only footprints" in order to preserve the natural and historical sites we visit. But the societal footprint we are leaving on Earth will win us no prizes for environmental stewardship. Species are disappearing from the global ecosphere at rates about a thousand times faster than only a millennium ago. A 2008 status report on the world's 5,487 known species of mammals showed that more than one fifth face extinction, and more than half show declining populations, largely because of declining habitat and hunting on the land, and overfishing, collisions with boats and ships, and pollution in the sea.[17] Gus Speth, the dean of the Yale School of Forestry and Environmental Studies, says that "Earth has not seen such a spasm of extinction since 65 million years ago," when the dinosaurs and many other species disappeared following the collision of an asteroid with Earth.[18]

We humans, in addition to manufacturing chemicals that impact the land, air, and waters of Earth, also produce heat, sound, and light, each with its own environmental consequences. It is well known that the buildings, roads, and roofs of a city absorb the heat of the day, and reradiate it through the night, in amounts greater than do the surrounding natural areas. Even the ground beneath the city has warmed substantially, as the warm footprint of heated buildings has replaced the cold air of winter at the ground surface. I recall that when I first started research into underground temperatures, my colleagues and I drilled a borehole near the laboratory for testing our instruments. The very first measurements of the temperatures down this hole surprised us—they showed

17. J. Schipper et al., "The Status of the World's Land and Marine Mammals: Diversity, Threat, and Knowledge," *Science* 322 (2008): 225–30.
18. James Gustave Speth, "Environmental Failure: A Case for a New Green Politics," *Yale Environment* 360 (2008), accessed at http://www.e360.yale.edu/content/print.msp?id=2075.

that the soil next to the building had warmed almost ten Fahrenheit degrees since the building was constructed many decades earlier.

If the gradual warming of cities is a subtle manifestation of urbanization, the generation of light is an obvious one. Driving at night on the highways of America, cities glow on the horizon like giant navigational beacons. And air passengers flying cross-country on a clear night see an illuminated panorama of the cities, towns, and isolated rural homes. The activities of people extend well into the hours of natural darkness, all made possible by our relatively newfound ability to "domesticate" light, just as we learned to domesticate fire thousands of years ago. The increase in urban illumination has long been known to impair the work of astronomers, who find once-visible stars gradually disappearing in an ever-brighter sky. And it deprives urban dwellers of one of the most beautiful experiences of all—viewing the millions of stars in the nighttime sky, visible only in areas free of light pollution.

I recall camping in the middle of the Kalahari Desert while engaged in some geophysical fieldwork in Botswana. In the dark of night—and it was truly dark—I found nothing so exquisite as gazing upward to see the majesty of a sky filled with uncountable stars, distant points of illumination in every direction, as far as the eye could see. In the Book of Genesis, God declares that he will "multiply your descendants into countless thousands and millions, like the stars above you in the sky." What a tragedy that so many people live their entire lives without ever experiencing this image.

But light pollution affects more than astronomers and city dwellers.[19] The other living creatures of the night, those we call nocturnal, display evolutionary behavior keyed to the existence of darkness. Some bats have become urbanized because their favorite insects swarm around urban

---

19. Verlyn Klinkenborg, an occasional columnist for the *New York Times*, provides a brief but very good discussion of light pollution in "The End of Night: Why We Need Darkness," *National Geographic,* November 2008.

light sources. And nocturnal mammals—many rodents, badgers, and possums—have become more vulnerable to predators because of their increased visibility at night. Some species are tuned to the longer-term variations in light and darkness that accompany seasonality. Many birds breed when daylight reaches a certain duration. If the nights are shortened artificially, and the days thereby apparently lengthened, the birds mistakenly think it is time to breed. Unfortunately, many of the grubs and insects that will nourish the hatchlings have not received the message that they should also advance their breeding cycle.

The biological rhythms of many species are tied to the daily cycles of light and darkness—these are called circadian rhythms. Because daylight provides the setting for work, the human body itself has evolved to use nighttime for sleeping. Alaskan, Scandinavian, and Russian hotels provide dark window shades to simulate darkness for summer tourists. Some geologists working summers at polar latitudes with twenty-four hours of daylight find that there is a real tendency to work too long, which leads to inadvertent fatigue and a greater susceptibility to accidents and illness.

No one needs to be reminded that populated areas are noisy areas: the sounds of vehicles throughout the day and night; of construction equipment digging holes, driving piles, and moving earth; of airplanes passing overhead; of motorboats, jet skis, and snowmobiles; of loudspeakers, radios, and televisions blaring. We go into the woods or the countryside for "a little peace and quiet," to leave behind the noisy urban environment.

But it may surprise you to learn that the waters of the oceans are not quiet sanctuaries free of anthropogenic noise. To be sure, the cacophony of the cities is absent, but sound travels very efficiently through ocean water, so that no place beneath the surface of the oceans is free of industrial sound. The cranes and conveyors that load and unload maritime cargo, the clanking of anchor chains dropping or lifting, the hum of diesel engines, the slow churning of massive propellers pushing ships through the sea, the pinging of sonar depth-finder systems, the underwater air guns and explosives used in seismic exploration of the sea bottom, naval training exercises with depth charges—the list of man-made sounds in the ocean is virtually endless. There is no place in the oceans—no place at all—where a sensitive hydrophone on the seafloor will not detect sounds of human origin.

Because so much of the ocean below the surface is dark, marine creatures often rely on sound for communication and navigation. The human noises have become distracting and disorienting, and in some cases of nearby noise, physically debilitating. Migration routes and breeding habits of marine mammals have changed in response to the geography and intensity of the noise—they, too, seek peace and quiet in the ocean. The mass grounding of whales and dolphins has on occasion been linked to high-intensity military sonar exercises in the vicinity. The U.S. Navy has acknowledged the "side effects" of this activity, but when challenged in court to end the noise-making, the navy argued that the necessity of conducting such exercises overrode the damage to the marine mammal population. In late 2008, the United States Supreme Court decided in favor of the navy.[20]

A BLANKET IN THE ATMOSPHERE

The parade of human effects on the land and water that we have just viewed certainly has left big footprints. But the biggest driver of climate change today, without question, is the impact of human industrial activity on the chemistry of the atmosphere. Since the beginning of the industrial revolution in the eighteenth century, there has been an ever-increasing extraction and combustion of coal, petroleum, and natural gas. The burning of these carbon-based fuels, previously sequestered

20. "Natural Resources Defense Council vs. Winter," *New York Times*, November 13, 2008.

in geological formations for millions of years, has rapidly pumped great quantities of greenhouse gases into the atmosphere, gases that affect Earth's climate by absorbing infrared radiation trying to escape from Earth's surface, thereby warming the atmosphere.

The industrial pollution of the atmosphere actually began long before the combustion of fossil fuels. Ice cores in Greenland show deposition of lead during Roman times, when that metal was used widely in the plumbing systems of the Roman Empire. The word *plumbing* derives from the Latin word for lead—*plumbum*—and Pb is the symbol for lead in the periodic table of elements. The smelting of lead and its use in manufacturing created lead dust that circulated widely in the atmosphere, some of which fell on Greenland and was incorporated into the accumulating ice. The deposition of Roman lead in Greenland came and went in tandem with the rise and fall of the Roman Empire. Lead reappeared in the Greenland ice when leaded gasoline made its debut as an automobile fuel, and largely disappeared when it was phased out as a fuel additive.

Another example of industrial atmospheric pollution is tied to the chemicals known as chlorofluorocarbons (CFCs). These inert nontoxic synthetic chemicals were first developed in 1927 as refrigerants, to replace the more combustible and toxic chemicals such as ammonia then in common use in refrigerators. In the decades following World War II, more and more uses for CFCs were discovered—for air-conditioning in homes, commercial buildings, and autos, and in the fabrication of electronics, foam insulation, and aerosol propellants. But the very property that made the CFCs attractive, their relative lack of reactivity with other chemicals, also made them very durable. Over decades they accumulated in the atmosphere, where they played an important role in the destruction of stratospheric ozone and the opening of the ozone hole in the last two decades of the twentieth century.

Ever since the beginning of the industrial revolution, air pollution has had deleterious environmental impacts, not the least of which have been the effects on public health. In 1948, a five-day-long incident befell the industrial town of Donora, Pennsylvania, on the Monongahela River, some twenty miles southeast of Pittsburgh. In what has been described as "one of the worst air pollution disasters in the nation's history,"[21] a combination of unusual atmospheric conditions compounded by smoke-stack discharges of sulfuric acid, nitrogen dioxide, and fluorine from two large steel production facilities led to a stagnant yellowish acrid smog. Respiratory distress for several thousand residents, and death for at least twenty, ensued.

In 1980, well before the fall of the Iron Curtain that separated the socialist countries of Eastern Europe from their Western European neighbors, I was invited to lecture in Czechoslovakia and East Germany about my geothermal research. On the highway from Prague to Leipzig through some nicely forested areas, I encountered a ten-to-fifteen-mile-long stretch of dead trees—defoliated gray matchsticks pointing mutely to the sky. Soon the cause became apparent: a massive chemical factory spewing great clouds of toxic pollution into the atmosphere, snuffing the life out of the forest downwind. The death of the forest was apparently considered acceptable collateral damage. One can only imagine what the airborne pollution did to the people and wildlife living nearby.

The combustion of coal has long been tied to health problems. In the United Kingdom, well known for dense, damp, chilly fogs, the addition of coal dust and tars to the fog forms a toxic brew associated with dramatic increases in pulmonary disorders. In the last decades of the nineteenth century, several "smog events" were accompanied by death rates as much as 40 percent higher than the seasonal norms. The infamous five-day London smog in December of 1952, an event that darkened the city at midday, was the product of a cold, dense fog made worse by increased burning of high-sulfur coal by London's chilly residents. It

21. *New York Times*, November 1, 2008.

led to more than four thousand coincident fatalities, with another eight thousand to follow in the weeks and months thereafter.

Industrialization and atmospheric pollution usually grow together. Following the rapid industrialization of Asia in the past few decades, severe atmospheric pollution is now commonplace over large parts of Asia. Dense brown clouds of industrial haze regularly blanket many large Asian cities, where automobiles and coal-fired power plants have grown, on a per capita basis, even faster than the population. In rural areas where consumerism has yet to penetrate, the simple aspects of daily living, such as cooking fires fueled by wood or dung, or the seasonal burning of the fields to prepare them for the next planting, also contribute to the haze that dims the Sun, retards agricultural production, and burns the lungs of millions of urban dwellers. The non-Asian world became aware of this acute pollution as China attempted to "clear the air" for the 2008 Olympic Games in Beijing, by closing factories and restricting automobile usage for weeks in advance of the Olympics.

The rapid post–World War II growth of coal-fired electric power plants in the central industrial states of America—Illinois, Indiana, Michigan, Wisconsin, and Ohio—was also followed by widespread atmospheric pollution that produced what came to be known as acid rain. The combustion of coal with high sulfur content yielded oxides of both sulfur and nitrogen, which reacted with water vapor in the atmosphere to produce potent acids. These acids were then deposited by rain and snowfall on the forests, lakes, and soil hundreds of miles downwind of the midwestern power plants, principally in the northeastern states and adjacent Canada.

In only a half century, the acid rain and snow led to a decline in the pine forests of the Northeast, and increased the acidity of some lakes in the Adirondacks to a level where fish eggs could not hatch. In the late 1960s and early 1970s, dead lakes surrounded by declining forests and acidic soils appeared with increasing frequency. When the cause of the acid rain was recognized, it provided some of the motivation for passage of the Clean Air Act in 1970, which, among other things, led to installation of scrubbers on smokestacks to capture the sulfur and nitrogen compounds that caused the downwind acid rain. The authors of the Clean Air Act, however, did not foresee a more dramatic environmental consequence of the burning of coal—the production of CO_2 that would accumulate in the atmosphere and dissolve in the oceans, altering Earth's climate and producing a more acidic ocean. Let us now take a close look at this unanticipated world-changing pollutant.

CO_2

In 1958, Charles David Keeling of the Scripps Institution of Oceanography in California began making measurements of the CO_2 in the atmosphere atop Mauna Loa, in Hawaii. These daily measurements, which have continued to the present day, provide the world's longest instrumental record of the direct consequence of burning carbon-based fossil fuels. This continued monitoring of carbon dioxide levels in the atmosphere is another example of the important Cinderella science that I describe in chapter 4.

These measurements show an increase in CO_2 concentrations in the atmosphere from 315 parts per million (ppm)[22] in 1958 to 390 ppm in 2009, an increase of 22 percent over just the past half century.

Riding along on this upward trend of CO_2 in the atmosphere is a fascinating annual oscillation—a small seasonal up-and-down that is the signature of photosynthesis occurring in the plant life of Earth each year. Most of the land area on Earth, and therefore most of the land plants, are in the Northern Hemisphere. During each Northern Hemisphere summer the growth of plant life draws CO_2 out of the atmosphere; in winter, when much of the vegetation is dormant, the CO_2 resumes its

22. A concentration reported as a few hundred parts per million can be visualized by a big bag of rice containing a million grains—nearly all are white, but a few hundred are black. A representative sample containing a million molecules of Earth's atmosphere will have a few hundred molecules of CO_2.

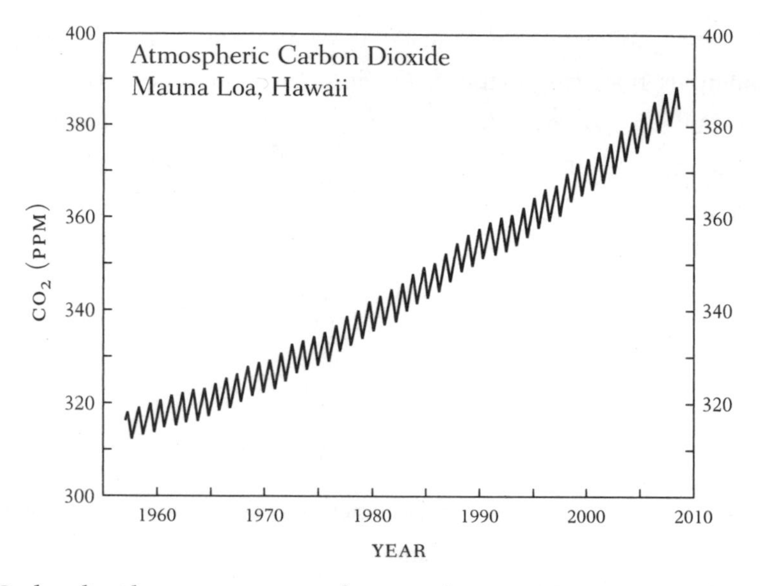

Carbon dioxide concentrations in the atmosphere over Mauna Loa from 1958 to 2008. Data from Scripps Institution of Oceanography

upward climb. This oscillation represents the actual metabolism of plant life on our planet, the annual "breathing" of Earth's green vegetation.

During the 1990s, CO_2 continued its climb above preindustrial levels at a rate of 2.7 percent per year, a rate more than twice as great as when Keeling began making measurements in 1958. In the first decade of the twenty-first century the rate has increased to 3.5 percent. Now, more than a half century long, the Keeling graph is the iconic signature of human energy consumption,[23] acknowledged even by skeptics.

The growth of CO_2 in the atmosphere is only part of the story—more than a third of all the CO_2 emitted by the global industrial economy enters the ocean, with effects that we are only beginning to understand. Oceanic uptake of CO_2 leads to a progressive decline in the pH of the water in the oceans, an indicator that the water is trending toward

23. Ralph F. Keeling, "Recording Earth's Vital Signs," *Science* 319 (2008): 1771–72.

becoming a weak acid, a process sometimes referred to as ocean acidification, although the seawater is not yet an acid. A lowering of the pH does result in a decline in the production and stability of minerals that calcifying organisms use to build reefs, shells, and skeletons, an effect likened to the onset of osteoporosis in the marine environment. This is already being observed along the Great Barrier Reef of Australia, where the calcification rate has decreased by about 14 percent since 1990, the largest decline in the past four hundred years.[24]

IN PERSPECTIVE

We can place the Keeling measurements into a much longer temporal context provided by microscopic bubbles of air trapped in ice. In chapter 3, while discussing the rhythms and orbital pacemakers of the last several ice ages, I mention that the deep ice core drilled at the Russian Vostok Station in East Antarctica revealed a 100,000-year periodicity in the temperature of precipitation of the annual snowfall. This same ice core, which at its bottom comprises ice as old as 450,000 years, also gives us a remarkable historical record of atmospheric carbon dioxide and methane.

The mechanism of record-keeping in the ice is fascinating—when snowflakes fall they accumulate into a very fluffy layer of flakes and air, which gets compressed into ice by the weight of subsequent snowfalls accumulating above. The air that is present becomes trapped as microscopic bubbles in the ice layer, and those bubbles constitute samples of the atmosphere at the time the snow recrystallized into an ice layer. The gases contained in the bubbles can be extracted and chemically analyzed to reveal the year-by-year greenhouse gas concentrations in the atmosphere over the complete time span represented in the ice core.

24. G. De'ath, J. M. Lough, and K. E. Fabricius, "Declining Coral Calcification on the Great Barrier Reef," *Science* 323 (January 2, 2009): 116–19.

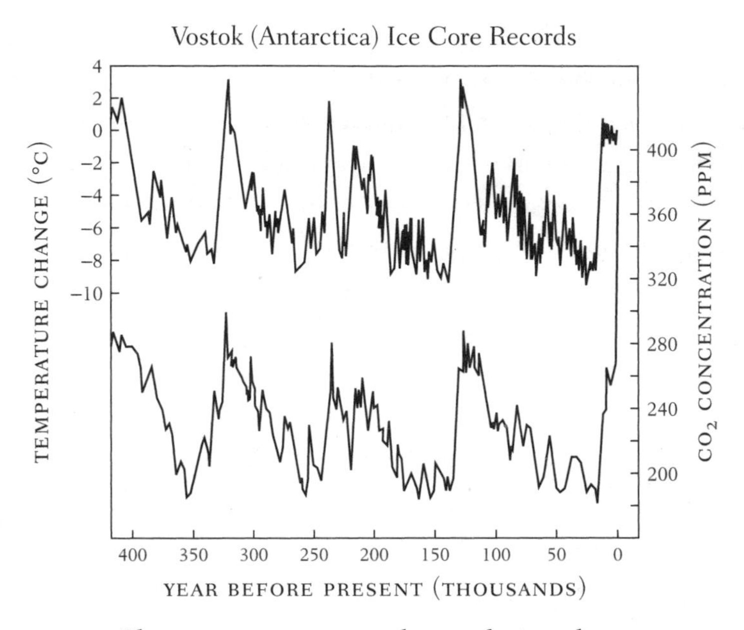

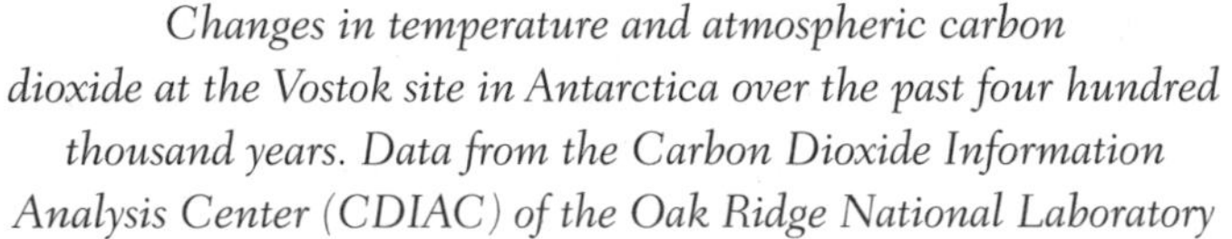

Changes in temperature and atmospheric carbon dioxide at the Vostok site in Antarctica over the past four hundred thousand years. Data from the Carbon Dioxide Information Analysis Center (CDIAC) of the Oak Ridge National Laboratory

These remarkable graphs show not only that the temperature went up and down with a 100,000-year periodicity, but that CO_2 did as well, in a pattern that is almost a direct overlay of the temperature record. The fundamental lesson of this overlay is that temperature and CO_2 are closely related in the processes that produce ice ages. A second observation drawn from these graphs is that atmospheric CO_2, over the course of the four complete ice age cycles shown, ranged roughly between 200 and 300 parts per million (ppm), from the cold of a glacial maximum when ice sheets covered vast expanses of the northern continents, to the briefer warm times that separated glaciations. Ice cores from elsewhere in Antarctica have extended the history of temperature and CO_2 back some 800,000 years, but throughout this long record CO_2 did not move out of the 200- to 300-ppm range.

But no longer. At the beginning of the industrial revolution in the middle of the eighteenth century, ice bubbles showed a concentration of CO_2 at around 280 ppm, near the upper value of 300 ppm that characterized the earlier interglacial periods. By 1958, when Dave Keeling began his measurements of atmospheric CO_2 atop Mauna Loa in Hawaii, it had already reached 315 ppm, well beyond the upper limit of the past 800,000 years. The concentration in 2009, a half century after Keeling began these measurements, reached 390 ppm, and is increasing by 2 to 3 ppm each year. It will likely cross the 400 ppm threshold in just a few more years. In the absence of any effective mitigation of greenhouse gas emissions in the near future, CO_2 in the atmosphere will reach the 450 ppm mark by 2030, a level that most climate scientists think will be accompanied by increasingly dangerous changes in our climate, which I describe in the next chapter.

In retrospect, a landmark point was reached in the mid–twentieth century when the concentration of CO_2 and other greenhouse gases moved out of the range of natural variability displayed over the previous 800,000 years. And it has become clear that yet another tipping point had been crossed, a point in time when the human influence on climate has overtaken the natural factors that had previously governed climate. Since then, population growth and increasing technological prowess have made the human influence increasingly dominant. Today the atmospheric concentrations of carbon dioxide and methane are substantially above their respective preindustrial levels, due to the contributions by humans and our many machines. This change in atmospheric chemistry is the real signature of human industrial activity, and the most important driver of contemporary global warming.

THE PRINCIPAL DIVISIONS of the geological time scale—such as the Paleozoic, Mesozoic, and Cenozoic eras, which together span the last 570 million years or so of Earth's history—have beginnings and ends marked by major changes in the character and distribution of life on Earth. Major extinctions on many branches of the tree of life mark the boundaries between these eras. Trilobites and many other marine invertebrates thrived in the Paleozoic, but did not survive into the Mesozoic. The Mesozoic era was the age of reptiles, with the dinosaurs the prime megafauna. But the saurians and many other taxa met their demise as the result of an asteroid impact sixty-five million years ago.

The Cenozoic is known as the mammalian era, and many of its subdivisions are associated with major climatic events: the temporal boundary between the Paleocene and Eocene was a very warm interval, likely caused by releases of the strong greenhouse gas methane from the ocean floor into the atmosphere. The Pleistocene is the period encompassing the most recent ice ages, followed by the Holocene, representing the eleven thousand years since the end of the last ice age. During the Holocene the climate has been remarkably steady, and *Homo sapiens*, that amazingly capable mammal, has flourished.

For nearly three million years, representatives of the genus *Homo*, both ancient and modern, have been passengers aboard Earth on its annual journey around the Sun. For most of that time, humans were but pawns—albeit increasingly clever ones—in the hands of nature, adapting to changes in climate and food and water as best they could. But in the past few centuries, after we domesticated fossil energy to amplify our personal strength, we humans are no longer simply passive passengers on the planet—we now are the dominant species of the planet. Unwittingly, we have become the managers of the intricately intertwined Earth system of rock, water, air, and life—the lithosphere, hydrosphere, atmosphere, and biosphere. And we are belatedly discovering that we are clumsy managers, woefully unprepared for this endeavor, and undergoing rough-and-tumble on-the-job training.

As recounted in this chapter, humans are having a profound impact on the landscape, on the waters, and on the living things of Earth. People have plowed up large areas of the land surface, have moved soil and rock at a rate much greater than natural erosion processes do, have controlled the flow of nearly every major river in the world, and have significantly altered the chemistry of lakes, rivers, the atmosphere, and oceans. We humans have wrested control of the carbon cycle, the nitrogen cycle, and the hydrological cycle away from nature, and have overpowered all other life forms on Earth, driving many to extinction. These clear and deep human footprints signal a major change in life on Earth, a reshuffling of the deck of life, with we humans moving to the top of the deck in only the eleven thousand years of the Holocene. Might not this reordering of life on Earth, driven by human fecundity and ingenuity, qualify for a new epoch in the geological time scale?

THE ANTHROPOCENE

Some say yes, and have given this epoch[25] a provisional name: the Anthropocene.[26] This name was first proposed informally by Paul Crutzen and Eugene Stoermer,[27] although several earlier authors have elaborated on the concept of a human dominated period in Earth's history. In a short essay titled "Geology of Mankind,"[28] Crutzen, a co-winner of the 1995 Nobel Prize in chemistry for his work on the chemistry of ozone depletion, again argued that humans have become the dominant species of the planet, and by the usual standards for defining a new unit of the

---

25. A geological epoch is part of a hierarchy of geological time subdivisions. A geological era comprises several periods, which in turn comprise several epochs. For example, the Pleistocene epoch occupies the interval of time between 1.8 million years ago up to 11,000 years ago, and is part of the Neogene period (23 million years ago to the present), which in turn is part of the Cenozoic era (65 million years ago to the present).

26. *Anthropocene* etymologically derives from the Greek *anthropinos*, meaning "human," and *cene*, signifying "new" or "recent"; the latter root relates to the fact that the epoch is part of the Cenozoic era, the geological era comprising the most recent life, in contrast to the earlier Paleozoic and Mesozoic eras.

27. P. J. Crutzen and E. F. Stoermer, "The Anthropocene," *International Geosphere Biosphere Newsletter*, no. 41, Royal Swedish Academy of Sciences, Stockholm, 17–18.

28. P. J. Crutzen, *Nature* 415 (2002): 23.

geological time scale, that a new subdivision was warranted. Crutzen points to the late eighteenth century as the start of the Anthropocene, when fossil fuels began to drive James Watt's new steam engine and the concentration of carbon dioxide in the atmosphere began to increase. William Ruddiman, a climate scientist at the University of Virginia, places the beginning of the human influence on the terrestrial environment five thousand years earlier, when agriculture began to generate the greenhouse gas methane, and deforestation led to more carbon dioxide remaining in the atmosphere.[29]

By any measure, humans have moved to center stage. We have placed our footprints indelibly upon Earth, and by changing the chemistry of Earth's atmosphere, we have inadvertently begun a planet-wide experiment with the global climate. In the next chapter I examine the consequences of this experiment, in terms both of changes that have already taken place on Earth, and of projected changes that we will face in the future. Today ice is already retreating because of human activities on Earth, and is perhaps on a trajectory to disappearance. Will future generations live on a world without ice?

29. W. F. Ruddiman, "The Anthropogenic Greenhouse Era Began Thousands of Years Ago," *Climatic Change* 61 (2003): 261–93.

CHAPTER 7

MELTING ICE, RISING SEAS

It is very difficult for someone living in the United States to grasp the fact that if the sea level rises just a few feet, a whole nation will disappear.

—BEN GRAHAM
Ambassador to the United States
from the Republic of the Marshall Islands

In the far reaches of the South Pacific Ocean sit many small islands—coral-fringed atolls that have formed on the subsiding calderas of extinct volcanoes. The growth of coral is fast enough to maintain the reef surface essentially at sea level, keeping pace with the slow geological subsidence of the ocean floor supporting the base of the volcano. Charles Darwin is best remembered for his compelling formulation of biological evolution, but he also was the first to recognize the role of volcanoes in the formation of coral atolls. Myriad low-lying islands in the South Pacific have been home to Polynesian, Melanesian, and Micronesian communities that have populated the islands for several thousands of years. These islanders have survived many challenges, including brutal occupation and warfare during World War II. But in the face of their newest challenge—rising sea levels from a warmer climate—these communities have no high ground to retreat to. Their only option is evacuation, mostly to places culturally alien to them. As Ambassador Graham says, it is indeed very difficult, and not only for Americans, to grasp the fact that if the sea level rises just a few feet, whole nations will disappear.

Tuvalu is a nation of twelve thousand people living on the atolls of the Ellice Islands, some two thousand miles north of New Zealand. Most of the islands sit only a few feet above sea level. Funafuti, the principal atoll, has an airstrip built by the United States during World War II. Today that lone runway provides the only easy connection with neighbors in Fiji and Samoa. But the airstrip has become increasingly vulnerable to partial inundation at the time of very high tides.[1] Tidewater in an atoll does not just move inland from the shoreline—it seeps upward through the coral and soil from below. The slow rise of sea level due to climate change has given the tides a head start, so that at some time in this century even ordinary tides will begin to force water to the surface to form shallow tidal lakes. Tuvaluans for a while longer will live on a saturated sponge that gets squeezed with regularity. But within this century they probably will have to abandon their home.

CHANGES IN THE ICE, water, landscape, and life of Earth go hand in hand with a change in its climate. In the broadest of terms, a shift is now under way in the balance between ice and water—ice is diminishing, water increasing. In the parlance of Earth's

1. For an interesting discussion of the Tuvalu flooding, see "A Sinking Feeling," by Samir S. Patel, in *Nature* 440 (April 6, 2006): 734–36.

hydrological cycle described in chapter 2, we are witnessing a transfer of H_2O from the solid cryosphere to the liquid hydrosphere. Mountain glaciers have been retreating, Arctic sea ice has been diminishing, the Greenland ice cap has been melting, permafrost has been shrinking, and sea level has been rising. And because of the continuing burning of fossil fuels, the greenhouse gas CO_2 continues to increase in the atmosphere, bringing yet further warming. What will be the consequences of this continuing climatic trend in the near future? The answers to this question differ from place to place, and from one elevation to another.

It is a mistake, however, to think that climate change is some abstract characteristic of the future. To the contrary: changes in the climate have been taking place for decades, and are continuing into the future at an ever faster pace. Indeed, the many changes already observed in the natural world, along with the millions of temperature measurements in the atmosphere, oceans, and rocks, are what persuaded the IPCC to conclude in 2007 that the warming of Earth is unequivocal.

In most of the continental mid-latitudes, including the lower slopes of mountains, snow and ice make only seasonal appearances. Permanent snow and ice on mountains, of course, depends on where the mountain is located—on the Antarctic Peninsula permanent snow and ice begins at sea level, but in the contiguous states of America year-round mountain snowpack and glaciers are found only at high elevations—well above ten thousand feet—in Glacier and Rocky Mountain national parks, and atop Mounts Rainier and Olympus in Washington. In the polar regions, ice dominates the landscape at all elevations year-round.

As snow cover lessens and glacial ice melts, it will not be just the scenery that changes. Water for municipal systems and agriculture in the foothills and plains surrounding high mountains comes from melting of both the annual snowpack and much older glacial ice, residual from colder times of the past. This is the very water that millions of people drink from the tap, that flushes sewage from towns and cities, and that waters the crops in the fields. Meltwater derived from this snow and ice is well timed for agricultural purposes, coming in the spring planting and summer growing seasons. But in a warmer world, where instead of snow, precipitation comes as rain that runs off when it falls, the water is not stored for later delivery during the agricultural season. And when the glacial ice of the mountains is finally gone, that source of water will disappear forever.

SEASONAL SNOW AND ICE

Let us begin a tour of the diminishing domain of snow and ice at its most tenuous geographic margins, at those latitudes where snow covers the ground for only a few days at least once a year. Where wintertime temperatures are already hovering around the freezing point, a little warming will mean the end of an already short period of snow cover. This occurs not just because there are fewer days sufficiently cold for snow to fall, but also because the dark peripheral ground, covered less and less by snow, progressively absorbs more heat over the year, and impedes the snow that does fall from accumulating. The southern margin of annual snow occurrence in the United States and Europe is slowly shifting northward a few miles each decade. In the mid-twentieth century some winter snow in Memphis was never welcome, but not uncommon. Now, in the early twenty-first century, a snowless winter in Memphis is not so rare, but ice storms from freezing rain, even less welcome than snow, are more frequent. Over the past fifty years the area covered by snow in North America has diminished in all months except November and December. February used to be the month of maximum snow cover, but that honor now belongs to January.

Wherever precipitation is delicately balanced at the freezing point, freezing rain is just as likely as snow. As a result, the incidence of ice storms is at a maximum, along with the special inconveniences of downed tree branches and power lines, and glazed roadways that promote fender-bender collisions. And as the number of days when the temperature is below freezing declines and snow falls less frequently, there are adverse consequences for outdoor winter sports. Nowhere is this change in the snow regime being felt more than in the Alpine towns and villages of Switzerland, where the local economies are heavily dependent on winter tourism. In mountainous terrain the snowline is creeping upward about seventy feet per decade. Studies of snow depth and duration[2] over the past sixty years show that a regime shift began in the late 1980s, when snow days declined by 20 to 60 percent. The reduction in snowfall coincided with an upward shift in the average wintertime temperatures, leaving little doubt about what was behind the diminished snowfall. The higher temperatures and the lesser amounts of snow have continued right through the first decade of the twenty-first century. Some Swiss ski slopes at vulnerable elevations are trying to preserve their wintertime snow and, in some cases the glacial ice beneath, with summertime plastic covers to shield them from the Sun. Ski resorts that make snow artificially because they are already challenged by inadequate natural snow are using their snowmakers more frequently, as they face the financial stress of a shorter natural season.

The warming of the waters in the Great Lakes of North America has led to later freezing and earlier melting of lake ice, as well as a diminished area with ice cover. The shorter period and smaller area of ice cover has had the consequence of increased loss of water due to more evaporation from the open lakes in winter. This has caused a drop of several feet in the lake levels over the past two decades, to a point where the upper Great Lakes—Superior, Michigan, and Huron—are approaching record lows. Lakeside cottages once a few steps away from the shoreline now see a beach a half-mile wide in some places, with wetland vegetation taking hold. And the entry channels to some major ports have become so shallow that without frequent dredging, they become impassable to the big

2. Christoph Marty, "Regime Shift of Snow Days in Switzerland," *Geophysical Research Letters* 35 (2008): article L12501.

freighters on the Great Lakes. These thousand-foot-long behemoths, were they at sea, would be too large to pass through the Panama Canal. In 2007, five fully loaded cargo ships ran aground attempting to enter the harbors at Muskegon and Grand Haven on the eastern shore of Lake Michigan. After being tugged free, they had to go across the lake to Milwaukee to off-load some cargo, before returning—floating higher—to unload the rest. The equally costly alternative is to carry a lighter load at the outset.

FROM THE MOUNTAINS

Long before humans came to the high Sierra Nevada of California, snowmelt fed streams tumbling through quartz veins containing gold, eroding and transporting the precious metal downstream. Where the currents slowed, the gold was dropped in the sand and gravel of the streambeds, later to be discovered by nineteenth-century prospectors. Today, it is the water itself that is the treasure—the snowmelt provides much of the annual agricultural water for California's fertile Central Valley, which stretches from Sacramento to Bakersfield. The winter snows that have for years made Lake Tahoe a popular skiing destination and Squaw Valley the site of the 1960 Winter Olympics, undergo springtime melting to swell the Sacramento River and deliver water to thirsty vegetable and fruit farms in the Central Valley. A little farther south, the Merced River, with its source in the great glacially carved valleys of Yosemite, also heads downhill to irrigate another section of the Central Valley.

But what happens when more of the precipitation comes as rain instead of snow? That H_2O is not stored in the winter snowpack—it does not wait for springtime to begin the downward journey to the dryer expanse of the Central Valley and then onward to the sea. Storage reservoirs along the way do not have the capacity to simply hold the water until later. So the water arrives much earlier than needed to help grow the produce of spring and summer, and when the agricultural calendar does call for delivery, the water is less abundant. A late-summer soil moisture deficit is a common result.

The dams and storage reservoirs along the waterways descending from the high Sierra serve another purpose—hydroelectric power generation. When water arrives early at already full reservoirs, it must bypass the dams via spillways. But every drop of water that bypasses the dams also bypasses the electric generators, the result of which is a deficit in power generation to accompany the deficit in soil moisture. If water is held in the reservoirs longer to provide steady hydroelectric power, the downstream flow is in places inadequate for agriculture and to maintain aquatic habitat for spawning salmon. The challenges of water management in an already semi-arid region are many. Within the larger expanse of the United States, water resources are already overallocated among agriculture, urban needs, ecosystem maintenance, hydroelectric energy, and recreation, at a time when demands in each sector are increasing.

In Europe, the Rhine River is fed in part by melting snow in the Alps and in part by rainfall over the low-lying parts of the river basin. A warming climate is changing the discharge of the Rhine to a rainfall-dominated regime—one of increasing winter flow and decreasing summer flow. The longer and more frequent low-flow episodes of summer are already apparent: less water for households, industry, agriculture, river transportation, and hydroelectric power during the time of peak summer demand.

Warming also shortens the snow season at both ends—winter snows begin later, and spring melting begins earlier. Already the peak stream flow from melting snowpack is appearing earlier in the season; by mid-century it is projected to arrive a full month earlier than the historical norms in the western United States.[3] The seasonal shift in melting also leads to longer summers and longer dry seasons, with more opportunities for wildfires. Research on the wildfires in the American West

3. T. P. Barnett, J. C. Adam, and D. P. Lettenmaier, "Potential Impacts of a Warming Climate on Water Availability in Snow-Dominated Regions," *Nature* 438 (2005): 303–9.

has shown that an extended dry season translates into a more intense fire season, with more wildfires that burn longer.[4]

Contributing to the potential for wildfires are the large areas of forest succumbing to insect infestations in western North America. Vast areas of the pine forests in the western states and adjacent Canada are being decimated by the pine bark beetle.[5] In British Columbia more than thirty-three million acres have already been lost, an area about the size of the state of Louisiana, and the infestation has spread to Montana, Wyoming, and Colorado. The beetle has crossed the Continental Divide into Alberta and is now making an appearance in the forests around the Great Lakes. This beetle is no stranger to the pine forests—it is not a recent invasive species—but the damage to the forests by it has grown dramatically in recent decades. What has happened? In the past, long and deep wintertime freezes controlled the bark beetle population, but now, with warmer winters and shorter deep freezes across the region, the beetle is suffering much less seasonal attrition. Greater numbers of hungry beetles are emerging each spring in search of nourishment, and the pine forests are their diet of choice. The dead trees then become fuel for wildfires.

Much higher on mid-latitude mountains, beyond the zone of seasonal snowpack, one encounters real glaciers, the streams of ice formed by compression and recrystallization of the snow of centuries past. In the contiguous United States these ice flows are few, principally atop Mount Rainier and other volcanic peaks in the Cascade Range of California, Oregon, and Washington, and in Glacier National Park in Montana. Most of these mid-latitude glaciers are remnants of the more extensive mountain glaciers of the last ice age. The extent of these former rivers of ice can be seen in the now-empty U-shaped valleys of the Sierra Nevada and Rocky mountains, as well as high up on the flanks of

4. A. L. Westerling et al., "Warming and Earlier Spring Increase Western U.S. Forest Wildfire Activity," *Science* 313, no. 5789 (2006): 940–43.
5. Jim Robbins, "Spread of Bark Beetles Kills Millions of Acres of Trees in West," *New York Times*, November 18, 2008.

the Cascade Range volcanoes. All of these remaining patches of ice are shrinking—within the current century, Glacier National Park will lose the very features that give it its name.

In Alaska, a score of long ice streams radiate downward from the twenty-thousand-foot peak of Mount McKinley, in Denali National Park, but surveys of the Denali glaciers show they are thinning and retreating rapidly. At lower elevations, along the coast of the Gulf of Alaska and in Glacier Bay National Park, the glaciers delivering ice to the sea have felt the effects of climate change as well. In Glacier Bay, modern cruise ships today sail into the bay and up the fjords to the glacier fronts. Only two centuries ago, when explorer George Vancouver visited the area, the bay was almost completely frozen over and virtually inaccessible.

Earth is losing ice today even faster than during the warming that followed the Little Ice Age of the seventeenth and eighteenth centuries. Projections for the future indicate that it will likely continue to do so, and at an accelerating pace. Every decade for the foreseeable future will see the loss of more ice. Vulnerable places, such as high mountains in equatorial and temperate latitudes, will see ice vanish soon.

The Andes mountain range forms an impressive spine running the full length of western South America. The high peaks of the Andes, with summits near twenty thousand feet, are also sites where ancient ice is present and is replenished with annual snowfall. But over much of the extent of this long mountain range, the mountains form a curtain that captures atmospheric moisture at high altitude, leading to a deficit of rainfall at lower elevations.

For some two thousand miles along the Pacific margin of South America there is a coastal desert, broken only by thin green ribbons where rivers and streams bring water from the high ice fields and snowpack. The villages, towns, and cities on the western slope of the Andes in Peru and Chile are made possible by the water rushing down from melting snow and ice above. Fields of agriculture—the abundant fruit and flowers and the remarkable vineyards of the region—are possible only because of the meltwater from the high Andes. La Paz, the administrative capital of Bolivia, draws much of its municipal water and all of its electricity from glacial melt. Lima, the capital of Peru, and its port city of Callao rely on snowpack and glacial meltwater to flush the municipal sewage (much of it untreated) to the sea.

But the warming of the climate is imperiling this source of water. Mountain glaciers from Peru to Patagonia are losing their ice to a warmer world, and are on the path to disappearance within a few decades. The Quelccaya ice cap in Peru, the Chacaltaya Glacier in Bolivia, Perito Moreno of Argentina (flowing eastward on the other side of the Andes), San Rafael Glacier in Chile, the Darwin glaciers flowing out of the South Patagonia ice field into the Beagle Channel—all show an unmistakable loss of ice at a rate even faster than occurred when the region emerged from the last ice age. The extent of Andean ice today is less than the region has known for at least five thousand years, and is undergoing attrition at a pace not seen since humans took up residence along the Andes fourteen thousand years ago. In 2009 the World Bank published a report projecting that the imminent loss of Andean glaciers will affect the water supply of nearly eighty million people and significantly reduce hydroelectric energy production in the region.

Across the Atlantic, glacial ice has a toehold even in equatorial Africa, atop nineteen-thousand-foot Mount Kilimanjaro, well known from Ernest Hemingway's "The Snows of Kilimanjaro." But the toehold is slipping fast—Kilimanjaro has seen a decline in its glacial cap throughout the twentieth century, and will likely be ice-free by 2020.

The Himalaya mountain range of Asia, comprising the long, high, remote boundary between India and Pakistan on the south and the Tibetan Plateau of China to the north, is often called "the roof of the world" for good reason. The highest peak, Mount Everest, stands above twenty-nine thousand feet, and more than one hundred others exceed twenty-three thousand feet. No mountain on any other continent reaches that elevation. In Sanskrit, the word *Himalaya* means "home of the snow," also for good reason. After Antarctica and the Arctic polar region, including Greenland, the ice mass on this Asian roof is the world's third largest. More than fifteen thousand glaciers flow downward from the high Himalayas, with a combined ice volume equivalent to some three thousand cubic miles of freshwater.

Each year these glaciers yield meltwater that provides a little less than half of the annual flow volume of the Brahmaputra, Ganges, and Indus rivers, the three principal rivers of South Asia that flow to the sea across Bangladesh, India, and Pakistan. The other half of the flow comes from the annual monsoon rains and snowmelt from the high mountains. But the contributions to these rivers are not spread evenly over the year—the monsoon dominates during the rainy season, and glacial meltwater is the principal source of water during the dry months.

A long-term warming of the atmosphere is well documented in the region, and measurements of the extent of glacial ice leave no doubt that the glaciers of the Himalayas are melting, and at an accelerating rate over the past two decades.[6] What message are these data delivering? Something akin to an urgent telegram that says that the volume of freshwater contained in the glacial ice of the Himalayas will last only another two to three decades. When the ice is largely depleted, the dry season flow of the Indus, Ganges, and Brahmaputra will diminish—already the lower Ganges is nearly empty for several months of the year. The lives of more than a billion people are intertwined with these rivers. Ironically, as the melting accelerates, there will be a temporary increase in water availability, until the shutoff arrives abruptly.

Ice on the Tibetan plateau north and east of the Himalaya also supplies the Irrawaddy and Mekong rivers flowing through Burma, Thailand, Cambodia, Laos, and Vietnam, and the Yellow and Yangtze rivers coursing through China. These rivers, too, face the loss of glacial meltwater in just a few decades hence. One of the principal glaciers feeding the Yangtze, the largest river in China, has retreated half a mile in a little

---

6. Natalie M. Kehrwald, Lonnie G. Thompson, et al., "Mass Loss on Himalayan Glacier Endangers Water Resources," *Geophysical Research Letters* 35 (2008): article no. L22503.

more than a decade. The loss of the glacial water has grave implications for agriculture, urban drinking water and sanitation systems, and hydroelectric power generation. Seasonal water stress will become a reality on all sides of the Himalayas, a fact of life, or death. Taken altogether, one quarter of Earth's population will within another decade be affected significantly by lesser snowfall and glacial ice loss. That number translates into almost two billion people—and most of them live in Asia.

Just as oil was the crucial resource of the twentieth century, water will be the prized resource of the twenty-first century. Already there is an international competition for water, and not just the decades-long tug-of-war between the United States and Mexico over Colorado River water. Nine nations share the Danube, six the Zambezi, four the Jordan. As politically charged a phrase as "the West Bank" is, one must remember that it is literally the west bank of the Jordan River. Water shortages are such a real threat for many nations that the potential for international conflict is also very real. In 2003 the Pentagon published a study of the implications of climate change for national security,[7] and pointed to water shortages as a special factor in international instability: "Military confrontation may be triggered by a desperate need for natural resources such as energy, food and water rather than by conflicts over ideology, religion, or national honor."

PERMAFROST

Farther north from the regions of simple seasonal snow cover the annual average temperature sits below the freezing point. There, aside from the summertime thawing of the upper foot or two—the "active zone"

7. P. Schwartz and D. Randall, "An Abrupt Climate Change Scenario and Its Implications for United States National Security," October 2003. http://www.greenpeace.org/international/press/reports/an-abrupt-climate-change-scena.

described in chapter 4—the ground is permanently frozen. This is the domain of permafrost. It extends over about 20 percent of Earth's land surface, mostly in the sub-Arctic and Arctic regions of Asia, North America, and Europe.

Earth's warming climate is taking a toll on permafrost terrains in many localities. As melting progresses to greater depths, the land surface suffers a disruption, stemming from the fact that ice occupies more volume than its equivalent melted water—the very same property that lets ice float on water. When deeper permafrost melts, the land surface above collapses into a jumbled irregular array of pits and hills. Houses and barns built on stable permanently frozen ground are undermined by ground subsidence; roads across the permafrost are broken up as if by an earthquake. Trees rooted in these slumping blocks are no longer vertical—their tilted orientation has earned them the description "drunken trees." The jumbled terrain that results from the melting of permafrost is called thermokarst. It takes its name from the true karst topography, the landscape caused by slightly acidic groundwater dissolving limestone and producing sinkholes and collapses over subsurface cave systems.

The Alaska pipeline is a long forty-eight-inch-diameter tube, a giant conduit through which oil produced at Prudhoe Bay, on the Arctic coast of Alaska, is transported to the tanker seaport of Valdez, some eight hundred miles south on the Gulf of Alaska. Along many stretches of this long traverse, the pipeline passes over permafrost. The design and construction of the pipeline in the 1970s took into account that warm oil, pumped from deep below Earth's surface, would melt the permafrost if the pipeline were buried in it. To avoid the possible buckling or breaking of a buried pipeline that melting permafrost might produce, engineers decided against burial. So for hundreds of miles the pipeline snakes across the tundra and through the boreal forest aboveground, perched on pedestals tall enough to let caribou wander unimpeded beneath.

Permafrost is something like concrete, in that the frozen water binds together soil, sand, gravel, and rock fragments. But when the permafrost melts, it releases the water and loses the cohesion that the ice provided. Rivers that drain into the Arctic Ocean—the Ob, Lena, Yenisei, and Kolyma—are showing increased flow volumes, and the Arctic Ocean is becoming less saline with the addition of the extra freshwater. The channels of the rivers themselves are becoming less stable, as formerly frozen riverbanks melt and collapse, releasing sediment that clogs channels and forces the rivers into new flow patterns that disrupt the already difficult challenges of life in the Arctic.

Melting permafrost, with its many sag ponds and exposed soil, is an important setting for the generation and release of gaseous methane to the atmosphere. The chemical composition of methane, CH_4, is as simple as that of carbon dioxide, CO_2, but methane is not a direct product of combustion. It is a fuel that is used for combustion, in kitchen stoves, in home furnaces and water heaters, in electrical generating plants, and indeed in some experimental automobiles. When methane is burned it produces CO_2 and, in the process, liberates heat.

Why should we be concerned about methane escaping from permafrost to the atmosphere? It is because methane is another greenhouse gas that plays a significant role in controlling Earth's climate. Once in the atmosphere, methane molecules—just like CO_2 molecules—trap infrared radiation leaving Earth's surface and warm the atmosphere. Methane is much less abundant in the atmosphere than CO_2—its concentration is less than 2 ppm compared to the 390 ppm of CO_2—but molecule for molecule methane is more than twenty times stronger in its infrared heat–trapping capacity than CO_2.

Sergei Zimov is a Russian scientist working in the far northeast of Siberia, in a small structure on the tundra with the innocuous designation Northeast Science Station, well north of the Arctic Circle, near where the Kolyma River empties into the Arctic Ocean. It is far from Moscow, about as far as Atlanta is from Anchorage, or Denver from Honolulu, and light-years away in terms of amenities. Winter temperatures are routinely −40° Fahrenheit, and the summer sky is clouded by mosquitoes. Recall that it was the mouth of the Kolyma River that Vitus Bering failed to reach in 1728 because of worries about being trapped in sea ice that begins to form there in September. Zimov and his wife, Galina, have been "trapped" there, voluntarily, for almost thirty years, observing the beginning of the end of the Arctic permafrost.[8] Their son, Nikita, was born there, and he is now part of the family scientific team, much like the three-generational Kozhov family team on Lake Baikal mentioned in chapter 4.

An area of western Siberia the size of Texas and California together has begun to melt from the top down. As another unheralded "Cinderella scientist" doggedly collecting scientific data, Zimov has recently been monitoring the methane bubbling up in the thaw lakes. The soil in regions of permafrost is effectively virgin and carbon-rich, the product of many episodes of tundra development between multiple glacial episodes over the past few million years. It is a very fertile soil for microbial generation of methane, now and in the distant past. Long inert in its frozen state, the methane has now been awakened by the warming of the Arctic. Zimov and other scientists who have come to join him on the tundra have found that the volume of methane and its rate of release are much greater than previously anticipated. And some of the methane is tens of thousands of years old, released from its long entrapment in the permafrost by the recent warming.[9]

Methane is also produced by other natural processes, in different environments and on shorter time scales. It escapes to the atmosphere from wetlands, peat bogs, and rice paddies. The common denominator of these botanical methanogenic locations is soggy organic soils, where microbial activity unites the carbon in the wet soil with hydrogen to

8. See the article by Adam Wolf about Sergei Zimov and the methane bubbling out of Siberia that appeared in *Stanford Magazine*, September/October 2008, 63–69.
9. K. M. Walter, S. A. Zimov, et al., "Methane Bubbling from Siberian Thaw Lakes as a Positive Feedback to Climate Warming," *Nature* 443 (2006): 71–75.

produce methane. It is a particularly effective process when the microbes are isolated from atmospheric oxygen—in places deep within the soil or in the muck at the bottom of lakes. Domestic herd animals also release methane at the terminus of their digestive tract, a process that in polite circles is referred to as "bovine flatulence." As dietary preferences around the world shift toward meat, this source of methane is growing.

Indeed, monitoring stations worldwide have shown a steady increase in atmospheric methane in the latter decades of the twentieth century, due both to increased rice cultivation and to the degrading permafrost. In the mid-1970s the methane reached double its preindustrial level, and today it stands at 2.7 times preindustrial. The rate of growth slowed briefly between 2000 and 2007, but has now resumed a rapid growth rate.[10] And the permafrost continues to degrade. Fortunately, it takes a long time for warming at the surface to penetrate to greater depths—permafrost five hundred feet down would be unaware of surface temperature changes over the past century, so slow is the downward propagation of thermal disturbances. The bad news, however, is that even if warming of the surface were to end today, the earlier warming would continue to diffuse into previously undisturbed permafrost. The unabated addition of methane to the atmosphere of course creates a positive feedback loop, accelerating the melting and release of more methane from the no-longer-permanent permafrost. And so this potent cycle of reinforcement continues to ratchet the warming of the Arctic upward.

ARCTIC SEA ICE

All along the northern margins of the degrading Arctic permafrost, in Scandinavia, Russia, Alaska, Canada, and Greenland, one encoun-

10. M. Rigby et al., "Renewed Growth of Atmospheric Methane," *Geophysical Research Letters* 35 (2008): article no. L22805.

ters the sea ice of the Arctic Ocean, described earlier in chapters 2 and 4. Historically, sea ice covered almost the entire ocean in winter, save for the occasional polynya.[11] Summertime saw some retreat, with about a third of the sea ice melting and breaking up. But large areas escaped breakup, and continued to block navigation. Explorers probing for pathways through the Arctic Ocean repeatedly met vast expanses of sea ice. For centuries the Arctic was virtually impenetrable to seagoing vessels.

Earth-orbiting satellites, launched in the late 1970s, offered the first synoptic views of the Arctic sea ice in its entirety. In earlier decades, the distribution of sea ice was gathered piecemeal from Arctic research stations, occasional areal photography, and from below the ice by American and Soviet submarines. These observations have shown that the summertime retreat and winter refreezing of sea ice were regular occurrences throughout the twentieth century, with little change in the extent of the summer ice loss. In the 1980s and 1990s, however, the summer melting and breakup began to consume more ice, until by the end of the twentieth century the summer sea ice extended over only 75 percent of the area it had covered at mid-century. Perennial ice, the older multi-year ice that survived the summer melting and breakup, was also diminishing at a rate of about 10 percent per decade. The winter refreezing replaced the older ice with first-year ice, and so the Arctic sea ice on average is becoming younger and thinner. The average thickness of the sea ice at the end of the century was reduced to about half its mid-century thickness. By 2007, summer sea ice was again down—to an area only 60 percent of its long-term average,[12] an all-time low since records have been kept.

11. A polynya, first mentioned in chapter 4, is an open area created and maintained by unusual combinations of wind and ocean currents.

12. J. C. Comiso et al., "Accelerated Decline in the Arctic Sea Ice Cover," *Geophysical Research Letters* 35 (2008): article no. L01703.

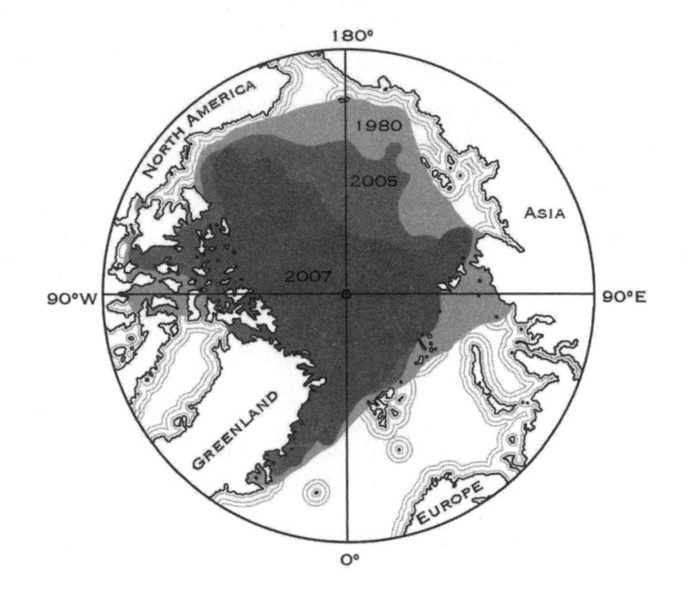

A map of the Arctic showing the reduction in summer sea ice that has taken place since 1980. Data from the University of Illinois, Urbana-Champaign

The area of sea ice lost since 1980 was half again greater than the entire area of the United States east of the Mississippi River. Taking into account changes in both area and thickness, almost 90 percent of the mass of summer sea ice over much of the twentieth century had disappeared by 2007. In 2006–7 the French ship *Tara* replicated the late-nineteenth-century drift of the Norwegian ship *Fram* across the Arctic Ocean, locked in sea ice. But *Tara* completed the drift in only five hundred days, compared to the three-year drift of *Fram*; the shorter drift time of *Tara* was attributed to the greater ease with which thinner sea ice can move across the ocean. In 2008, perhaps for the first time in many centuries, both the Northwest and Northeast passages were open at the same time.

The current rate of summer sea ice loss is exceeding all projections,[13] and there is a very real possibility that in only a few decades the Arctic Ocean may be ice-free in the summer, for the first time in fifty-five million years. Climate scientist Ian Howat of Ohio State University remarked that the loss of Arctic sea ice might be the largest change in Earth's surface that humans have ever observed.[14]

ARCTIC SEA ICE typically takes the brunt of punishing ocean waves and storms, but as the ice has all but disappeared in the late Arctic summer, the exposed coastline has become vulnerable to severe erosion. Villages in the Arctic have traditionally been built close to the sea to enable easy pursuit of seals, walruses, and fish. But the loss of shoreline protection has taken a toll on structures built near the sea. Shishmaref, Alaska, a small community with fewer than six hundred residents built on a small island in the Bering Strait, is facing the possibility of forced evacuation. Shishmaref has had human inhabitants for several thousand years, but it has always had the protection and convenience of surrounding sea ice during the storm season. In recent years, however, storms have time and again eroded the island. Shishmaref is literally losing ground, and the residents have had to relocate houses farther from the shoreline. Travel to the mainland for hunting moose and caribou, formerly easily accessible over the sea ice, has been impeded because the winter freeze of the sea has recently been delayed by as much as six weeks.

Just east of the Alaska-Canada border, along the coast of the Beaufort Sea, sits Tuktoyaktuk, a Canadian port town. The terrain of the area is low-lying, with little topographic relief. Sea ice formerly offered

13. Julienne Stroeve et al., "Arctic Sea Ice Decline: Faster Than Forecast," *Geophysical Research Letters* 34 (2007): article no. L09501.
14. Quoted in Alexandra Witze's "Losing Greenland," *Nature* 452 (2008): 798–802.

protection from big waves and storm surges, but today, erosion in Tuktoyaktuk is an increasing problem. As in Shishmaref, buildings are being abandoned to the encroaching sea. Multiple attempts at engineering shoreline protection have failed, in a losing battle with the advancing surf. Most of the coast surrounding the Arctic Ocean is increasingly vulnerable to destabilization, because the seawater accelerates the melting and destruction of the coastal permafrost and leads to the development of thermokarst. The temperature of seawater is clearly above the melting point of ice, and so contact between seawater and permafrost leads to rapid degradation.

It is not only the hunters on Shishmaref who have seen the loss of sea ice cut them off from large game. Polar bears are also affected. For years sea ice has served as a widespread platform for polar bears to move about in search of food, particularly the seals of the Arctic Ocean. Pregnant female bears spend winter in dens of snow and ice, emerging in spring after a half year without food, and with a new family of hungry cubs. An early breakup of the sea ice impedes their ability to hunt, forcing them to walk less and swim more in search of seals, a bad trade-off in terms of energy expended. Reproductive success is closely tied to the fitness of the mother—weakened females produce fewer and smaller cubs. Multidecadal studies of polar bears in the Hudson Bay area reveal declines both in average parental weight and in the number and weight of cubs. And the sea ice is breaking up a week earlier each decade.

Though stronger swimmers than the bears, the seals are faring no better. They, too, use the sea ice as a platform at calving time, and an early breakup of the ice leads to a premature separation of parents from the newborn, before the young seals are ready to face the world on their own. Walrus also rest on sea ice as they scavenge the shallow near-shore seafloor, but as the edge of the sea ice moves to deeper water, the walrus is faced with more arduous seafloor foraging.

MEANWHILE, AT THE BOTTOM OF THE WORLD . . .

Palmer Station, one of three U.S. research stations in Antarctica, was built on Anvers Island, along the Antarctic Peninsula, in 1968. It comprises a handful of prefabricated buildings that host research projects, a small boat anchorage, fuel storage tanks, and living and dining quarters for some forty people. The station, named for the early-nineteenth-century American sealer and explorer Nathaniel Palmer, is chiseled into a rocky outcrop, squeezed between the large Marr Glacier and the open sea. From the air the station is dwarfed by the vastness of the peninsular ice and the ocean that surround it. The research focus at Palmer is largely biological, but includes a seismographic installation and a meteorological station.

Since 1950, the Antarctic Peninsula has warmed by more than six Fahrenheit degrees, mostly due to a remarkable rise in wintertime temperatures of more than ten Fahrenheit degrees. This warming has reduced the extent of winter sea ice along the western edge of the peninsula by 40 percent and shortened the duration of sea ice cover there by almost three months. These changes in the sea ice along the peninsula are leading to changes in the ecology of the rich marine life of the region. But the changes begin not with the big animals—the whales, seals, and penguins—but rather at the bottom of the food chain, with the marine phytoplankton. Phytoplankton are single-cell plants that fuel their growth with sunlight. They flourish near the margin of the sea ice that grips much of the Antarctic Peninsula each winter. When spring arrives and the sea ice begins to break up, the phytoplankton bloom prolifically. They are then devoured by small shrimplike crustacean called krill, which are in turn the diet of choice for penguins and whales farther up the food chain.

If the warming trend of the peninsula continues, by 2050 midwinter temperatures will remain warmer than the freezing point of seawater, and sea ice will not form. Jim McClintock, a professor of polar and marine biology at the University of Alabama at Birmingham and a long-time researcher at Palmer, says that the warming will lead "to a regime change in the ecosystem. Krill are highly dependent on sea ice; without it they cannot complete their life cycle and breed successfully." The decline of sea ice in area and duration is particularly ominous for the krill, and for the penguins that eat them, and for the predatory seabirds that eat the penguin eggs and chicks, and so on, up the food chain. At the top of the list are the whales, which don't depend upon or even bother with the intermediate links in the food chain—they eat krill directly, by the ton every day.

One long-term study at Palmer has focused on the population dynamics of the three penguin species in the area—the Adélie, Gentoo, and Chinstrap. McClintock and colleagues Bill Fraser and Hugh Ducklow have pieced together the interdependencies of these penguins with the changing environment.[15] Because each species interacts with sea ice in different ways, principally in their access to food, the changes in the ice are driving the populations in different directions. Adélie penguins, the classic little tuxedo-clad seabirds, prefer wintertime foraging at offshore locations where nutrient-rich marine upwellings occur, sites made more accessible to them by the existence of widespread sea ice. By contrast, Chinstrap and Gentoo penguins prefer to forage along open shorelines. The declining sea ice has therefore been hard on the Adélies but generous to the Chinstraps and Gentoos.

I have been to Palmer Station four times over eighteen years. When I first visited in 1991 there was a big Adélie population on the islands around Palmer—noisy, smelly, coming, going. But on my most recent

15. James McClintock, Hugh Ducklow, and William Fraser, "Ecological Responses to Climate Change on the Antarctic Peninsula," *American Scientist* 96 (2008): 302–10.

Adélie penguins of Antarctica

stop, in early 2008, the Adélies were largely gone. Indeed, all along the northern half of the peninsula, Adélie populations are in decline. Deciphering these trends would not have been possible without year-in, year-out monitoring of the penguin populations, the sea ice extent and timing, and temperature and precipitation changes. The U.S. National Science Foundation has designated Palmer Station a Long-Term Ecological Research (LTER) site, a designation that assures funding for the long-term observational programs that track trends in the ecosystems. The NSF has established twenty-six LTERs, three of which are in polar regions—one on the north slope of Alaska and two in Antarctica.

ENCROACHMENT OF THE SEA

The most dramatic consequence of a world losing its ice will be the steady—and perhaps not so steady—rise of sea level. For the entire twentieth century and continuing into the twenty-first, sea level has been creeping upward, due to two principal causes. The first relates to some simple physics—seawater expands as ocean temperatures rise. The second is the return of meltwater to the ocean, after spending thousands of years as ice in temporary residence on the continents. Both causes lead to encroachment of the sea onto the land. There is already more water on Earth than can be accommodated in the deep ocean basins, and the excess has spread out to cover part of the continental shelves. As sea level continues to rise, foot by foot, shorelines will relocate inland.

The IPCC has estimated the likely sea level rise in the present century to be in the range of ten to thirty inches, which, when added to the eight-inch rise of the previous century, indicates that we are facing an increase of some two to three feet above the historic average sea level. Such numbers may seem modest, but in fact changes of that magnitude will have nontrivial consequences in shoreline erosion, infrastructure destruction, and population displacement. In making these estimates, the IPCC considered only the effects of continued thermal expansion of ocean water, the addition of new water by melting and runoff of continental ice, and the return of some groundwater to the sea. But the IPCC hedged their projections with one important caveat that I will return to shortly.

Low-lying places on Earth—those barely a foot or two above today's sea level—will be the first to feel the encroachment of the sea. The barrier islands and very gently sloping coastal plains along the Atlantic and Gulf coasts, many coral-fringed atolls of the Pacific, and major river deltas the world over are particularly vulnerable to a rising sea. These are the places that already take the brunt of storms and hurricanes. Almost every structure on the barrier islands at Galveston on the Texas Gulf coast was leveled by the winds and storm surge of Hurricane Ike in 2008, and as sea level creeps upward, these islands will be increasingly vulnerable to even lesser storms.

Along the Atlantic coast of the Carolinas, Georgia, and Florida there is a continuing struggle between the erosion and shifting sands of the barrier islands and the ever-increasing number of people who build houses on them. The houses are expensive, and so is the struggle—dredging sand to keep established channels open, trucking in sand to replace the beaches lost in storms. And there is always the temptation toward "hard stabilization"—the building of protective seawalls and other types of barriers—which almost always fail, leaving a massive jumble of concrete debris with bent steel reinforcing rods to degrade the landscape, in mute testimony to the futile attempt at forestalling nature.

Well before rising waters force abandonment of coastal and barrier island homes, the residents will also notice that their well water has turned briny, as a result of seawater pushing into coastal aquifers. Shortly thereafter sewage systems will begin to fail as seawater degrades bacterial action. Little of Florida from the Everglades south through the Keys sits more than three feet above sea level. During the storm surge of Hurricane Wilma in 2005, many of the Keys were under three feet of water and more than half the houses in Key West were flooded. Every part of this region is vulnerable to even a modest rise in sea level.

The levees that channel the Mississippi River through New Orleans are an example of hard stabilization. These mounds of earth and concrete walls are massive constructs designed and engineered by the U.S. Army Corps of Engineers. New Orleans does not sit a foot or two above sea level—much of the city is several feet below sea level. The life of New Orleans has for decades been entrusted to the levees, a trust that was betrayed during Hurricane Katrina in 2005. Katrina forced the evacuation of some 150,000 residents of New Orleans, and many never returned. Despite some post-Katrina repairs, there was another near-betrayal during Hurricane Gustav in 2008, when the storm surge lapped at the very top of New Orleans's floodwalls, splashing over with any slight provocation from the wind. Gustav was not the storm that Katrina was, but had sea level been only a foot higher, Gustav would have replicated what Katrina wrought. The slowly rising seas of the twenty-first century will make any hurricane a danger to New Orleans.

Venice is built on several islands in a marshy lagoon at the north end of the Adriatic Sea. The city proper and other occupied islands in the lagoon are home to one hundred thousand people. The canals of Venice lead to the open sea, and are only one indication that the city and the sea are like hand and glove. Piazza San Marco, perhaps Venice's premier tourist destination, is vulnerable to flooding from a multitude of causes—stiff winds from the sea, high tides, or intense rainfall. In early December of 2008 all three conspired to place St. Mark's thigh deep in water, a level within the range of the IPCC estimates of sea-level rise this century. With sea level creeping higher, Venice will be visited more frequently by flooding, even in the absence of a triple conspiracy. Authorities in Venice have long dreamed of building a flood barrier at the mouth of the city's lagoon to hold back the Adriatic at times of high tides and wind-driven water, but the costs to build the barrier have escalated faster than the available money. And meanwhile, sea level keeps rising.

Kiribati is an island nation in the south Pacific, with some ninety thousand citizens scattered on more than thirty atolls straddling the equator. Its principal community is on Tarawa atoll, the site of a ferocious struggle between the Japanese and Americans during World War II. Kiribati will be another of the countries that, like Tuvalu, will lose territory to rising sea level. Already many places are experiencing flooding, erosion, and saltwater contamination of the groundwater at the time of the twice-monthly high spring tide. In 2008, Kiribati asked Australia and New Zealand to open their doors to Kiribati citizens as permanent climate refugees. Kiribatians worry that no matter how the large industrial countries might attempt to prevent future climate change the oceans will continue to warm and sea level will continue to rise because of the long-lived greenhouse gases already in the atmosphere.

The Maldive Islands are an archipelago of twenty-six atolls in the Indian Ocean, south of the southern tip of India. Most are about three feet above today's sea level; none reaches more than seven feet. That topographic fact gives the Maldives the honor of being at the bottom of the list of nations ranked by their maximum elevation. Some 350,000 people reside in the Maldives, about one third in the capital city of Malé. Seawalls completely surround Malé for protection. Perhaps more prosperous than many island nations, the Maldives are thinking about creating an investment fund to buy land in India, Sri Lanka, or elsewhere as a destination for climate refugees. The Maldives recently experienced the devastation of higher seas when the tsunami associated with the 2004 Indonesian earthquake literally rolled over many of the islands.

THE DELTAS

Most rivers flow to the sea, completing the hydrological cycle by returning water evaporated earlier from the oceans. But big rivers also carry vast amounts of sediment eroded from the land. Deltas are geologically dynamic meeting places—where rivers and their sediment load meet the sea. When a river flows into the sea its current is slowed and the sediment settles. Over time the channel wanders back and forth over the accumulating pile of sediments to form a delta. The Greek letter delta (Δ) has the shape of a triangle, and is an appropriate descriptive term for the broad fan of sediment that rivers dump into the sea at their mouths. The deltas of Earth's major rivers—the Amazon, Ganges, Nile, Mississippi—extend over hundreds of miles, some parts slightly above sea level, others slightly below. Over time, some mud banks and sandbars will slowly sink below the surface, while other areas receive enough sediment to rise slightly above the sea and feel the sunshine for a few decades.

The Nile Delta meets the Mediterranean Sea along a 150-mile shoreline, about a hundred miles north of the apex of the delta where the river emerges onto the delta from its narrow floodplain after a long meandering journey across the Egyptian desert. Along the coast of the delta at its western edge sits the ancient and modern city of Alexandria, home to more than four million people. Much of the historic ancient city is now submerged beneath the sea, largely due to tectonic subsidence of the land associated with regional earthquakes. Alexandria's El Nouzha International Airport itself sits six feet below sea level, isolated from the sea by natural barriers.

The soil of the Nile Delta is very rich, and accordingly, rural parts of the Nile Delta are heavily agricultural. Over much of the history of the Nile, the sediment of the delta was replenished by the annual flood of the river, but since the completion of the Aswan High Dam in 1970, six hundred miles south of the delta, the annual flood no longer occurs along the lower Nile. The soil of the delta now requires fertilizers to maintain productivity. But the delta itself, no longer replenished with sediment from Africa's interior, is slowly retreating as shoreline erosion along the Mediterranean inexorably eats away at the delta's margin. And as seawater pushes its way into the subsurface aquifers, the groundwater in the delta is becoming more brackish. The erosion of agricultural land and the intrusion of saltwater will only accelerate as sea level rises due to climate change.

The Bengal region of eastern India and the adjacent country of Bangladesh sit astride the immense delta built by the Ganges and Brahmaputra rivers. These two rivers, both with headwaters in the Himalayas, drain a sizable fraction of Asian territory and deliver to the delta a heavy load of sediment eroded from the mountains. This area, with its large cities of Calcutta and Dhaka, has one of the world's highest population densities, on some of the lowest terrain. Calcutta, located some sixty miles from the Indian Ocean, has more than eight million residents, and its lower precincts are only five feet above sea level. Much of the coastal south of this huge delta is only a foot or two above sea level. It is a great oversimplification to speak of a coastline—the actual margin of the land with the ocean is like a giant frond, with bays and marshes extending far inland, separated by tongues of land that do not qualify for the adjective *dry*. Seasonally, about 10 percent of Bangladesh is underwater.

The terrain of the Ganges Delta is about as flat as one can find, rising only a foot above the sea for every twelve miles inland from the coast. By comparison, the Great Plains stretching across Kansas and Nebraska to the mountains of Colorado are a steep ramp, rising by about eight feet for each mile they sit westward of the Missouri River. A slope of one degree, which we might imagine to be a very small slope, rises ninety feet in a mile. A sea-level rise of a foot or two in this century—the conservative projection of the IPCC—will force a devastating displacement of millions of people living on the Ganges Delta.

The numbers are sobering—11,000 Tuvaluans, 60,000 Marshall Islanders, 90,000 Kiribatis, 100,000 Venetians, 150,000 people in the Big Easy, and millions in the deltas of the Nile and Ganges—all facing displacement by only a modest rise in sea level. A recent study[16] of where people live on Earth showed that 108 million humans, equivalent to the population of Mexico, live on terrain no more than three feet above sea level. Those numbers translate into 1.6 percent of Earth's population living on less than 1 percent of the land surface—that part of the land barely above sea level. In the United States along the Atlantic and Gulf coastal plain and barrier islands, more than 2.5 million people live no more than three feet above sea level.

Hurricane Katrina displaced about 150,000 people from New Orleans, and the 2008 California wildfires displaced almost 1 million—and both events severely tested our nation's ability to address the needs of environmental refugees. Cyclone Sidr, a November 2007 tropical storm in the Indian Ocean, killed more than 3,000 Bangladeshis and left more than one million homeless, with the full consequences of that population dislocation still to be determined. A three-foot rise in sea level this century, creating more than one hundred million climate refugees, is equivalent to about seven Katrina-scale evacuations every year,

16. R. J. Rowley et al., "Risk of Rising Sea Level to Population and Land Area," *Eos Transactions American Geophysical Union* 88, no. 9 (2007).

although much of the population displacement will occur later in this century.

In rich nations, much effort and treasure will be expended to forestall and avoid the advance of the sea. Levees will be strengthened, seawalls will be built, but most of the effort will be futile—it is almost impossible to hold back the sea along an extended coastline. The masters of that craft, the Dutch, already have recognized the near impossibility of the task, and have made plans to abandon some of their land to the rising sea. In poor nations, people will simply be forced to move to higher ground, ground already occupied by other people, who are unlikely to welcome the newcomers. If immigration now seems to be a thorny, controversial, and emotional problem in the United States and Europe, after three feet of sea-level rise and one hundred million climate refugees worldwide, today's immigration complexities will in retrospect look like a Sunday school picnic.

It is entirely possible that sea-level rise will not continue at the slow and steady rate we have experienced in the twentieth century. As seawater warms, its ability to take in CO_2 from the atmosphere diminishes because the solubility of CO_2 in seawater decreases as the temperature of the water increases. The likely result will be that in the present century, as the oceans continue to warm, they will be less of a sponge for atmospheric carbon dioxide. More CO_2 will remain in the atmosphere and enhance the heat-trapping greenhouse—thus accelerating the melting of land ice, and the warming and thermal expansion of the ocean water. And the physical mechanisms that cause water to expand are more effective at higher temperatures than at cooler temperatures. The physics and chemistry of the oceans and atmosphere make inescapable the conclusion that a warmer Earth will lead to sea level rising at an accelerating pace. Already the rate at which the sea was observed to rise in the period 1993–2003 exceeded the average rate over the longer period 1961–2003 by more than 50 percent.

THE THIRD TRENCH OF DENIAL

The climate contras have occupied several successive defensive trenches in their campaign of denial of anthropogenic climate change. The first trench was simply that of disputing the many observations that climate has been changing. The second was that if climate has been changing, the change has been due entirely to natural causes—that humans have had no essential role in climate change. Although some contras remain active along these lines of defense, many have retreated to a third trench: Who, they ask, will be unhappy with a CO_2-rich atmosphere and a warming world? They point out that CO_2 acts as a fertilizer, stimulating agriculture around the world, and that a warmer world will extend the growing season and enable more food production. And they muse about the pleasures of milder winters, free from the hassles of snow and ice. In their more rhapsodic moments they speak of a return to the Garden of Eden.

Indeed, the growing season is growing—in Alaska it is already two weeks longer than a half century ago. And it is true that CO_2 is a necessary ingredient for photosynthesis. But CO_2 alone is not sufficient for increased plant growth. Frequently in natural settings it is the availability of other nutrients—particularly water and nitrogen—that are the limits to additional biomass, not the availability of CO_2. Controlled experiments at Duke University's Duke Forest showed that elevated levels of CO_2 produced no incremental growth in trees rooted in nutrient-poor soil.[17]

Moreover, in nature, CO_2 does not vary in isolation. In a greenhouse-forced climate change, higher temperatures and changed geographic patterns of precipitation and soil moisture accompany an increase in

17. R. Oren et al., "Soil Fertility Limits Carbon Sequestration by Forest Ecosystems in a CO_2-enriched Atmosphere," *Nature* 411 (May 24, 2001): 469–72.

atmospheric CO_2. In one of nature's own experiments, the boreal forest in Alaska has not shown dramatic growth as the atmospheric CO_2 has increased because it is also experiencing heat stress from the warming climate. And the pine forests of western North America, far from flourishing in a higher CO_2 atmosphere, are suffering terribly from a resurgence of the pine bark beetle, a pest formerly kept under control by colder winters.

CO_2 is not a discriminating fertilizer—it fertilizes weeds as well as foodstuffs. And there is nothing to guarantee that the crop growth we might want to enhance will be able to outpace unwelcome invasive insects and plant species that also like the warmer and CO_2-richer atmosphere. If a greater use of herbicides and pesticides to control invasive species follows, the negative effects on public health and the general ecosystem will diminish whatever benefits might come from CO_2 fertilization.

The rhetorical question "Who wouldn't welcome warmer winters?" is both parochial and simplistic. It implicitly assumes that a warming that would ameliorate winter cold, snow, and ice in the latitudes where those phenomena occur would be welcomed also by the majority of Earth's population who do not live in that environment. Higher temperatures in already warm regions are not high on the wish list of those who live there. And of course a long-term trend toward warmer winters does not occur independently of changes in the other seasons—such a trend is often accompanied by a similar trend to warmer summers, with more heat waves and soil desiccation. The IPCC assessment of impacts accompanying a warming world shows that the tropics will fare poorly by almost any measure.

Even were there to be some regional benefits in the early stages of moderate climate change, as there well might be in some mid-latitude settings, those selective and limited benefits hardly outweigh the large-scale hardships and human displacement associated with even a modest rise in sea level. Higher seas will affect almost all of the world's coastlines adversely—large nations and small, rich nations and poor. There is no way that rising sea level can be described as beneficial.

THREE FEET OF SEA-LEVEL rise, however, may be just the beginning. The ongoing breakup of Antarctic ice shelves—the Larsen, Wilkins, and Ronne along the Antarctic Peninsula, and the Ross Ice Shelf farther around the continent to the west—has ominous implications for sea levels globally. These floating shelves of ice, in place for thousands of years, form buttresses that bottle up glacial ice on the land of the peninsula and West Antarctica, slowing spillage of ice into the sea. Those long-standing buttresses are now breaking away, and the glaciers are speeding up their delivery of land ice to the ocean.

When ice flows off the land and enters the sea for the first time, it raises sea level. It is nothing more than a large-scale replication of the process that takes place when you drop an ice cube into a glass of water. The submerged part of the floating cube displaces water and moves it aside, and up, in the glass. Once afloat in the glass, however, ice causes no further change in the water level as it melts. The same holds true for floating ice shelves—when they break up and melt, they do not change sea level because the effect of the floating ice on sea level occurred when the ice originally spilled out onto the sea. But the breakup of the shelves is an early warning sign, an icy "canary in the coal mine," because the breakup allows new land ice to spill into the sea, and the new ice does raise sea level.

The mass of ice bottled-up on the Antarctic Peninsula, were it to slide into the sea, would raise the global sea level about ten inches, roughly equivalent to the cumulative sea-level change observed during the twentieth century. However, the twentieth-century sea-level change was due principally to the warming and thermal expansion of the seawater globally, and to some new water added to the oceans from the melting of mountain glaciers and permafrost. Those were relatively slow processes in the twentieth century. But when ice is dropped directly into the ocean, the sea-level change takes place right away, before the ice melts. That is why an acceleration of ice loss from Antarctica and Greenland is worrisome—sea-level changes can come much more quickly than the relatively slow increases arising from warming alone.

The amount of ice on each of the Greenland and West Antarctic landmasses is equivalent to about twenty feet of sea-level change, and that on East Antarctica, about two hundred feet. Because much of the ice surface on East Antarctica is remote from the sea and very high, the temperature is extremely cold. It averages about −50° to −60° Fahrenheit over the year, and never gets close to melting, even in summer. Of all the ice on the globe, that in East Antarctica is thought to be the most stable. But Greenland and West Antarctica present a more ominous scenario.

GREENLAND

As the Arctic has warmed, summertime melting has been creeping to higher elevations on the Greenland ice cap. Some of the meltwater runs to the sea, and some ponds up in slight depressions to form blue lakes on the surface of the white ice. Occasionally these meltwater lakes disappear suddenly, draining into large fissures called moulins, which apparently penetrate the entire mile-thick ice sheet. Meltwater plunges to the base of the ice and makes its way along the interface between the rock and ice, much like subterranean rivers that flow deep within limestone caves.

The consequence of this deep subglacial flow is profound—the water lubricates the base of the ice sheet, allowing the ice to slip toward the sea at a faster rate.[18] The Jakobshavn Glacier, on the western coast

18. H. J. Zwally et al., "Surface Melt–Induced Acceleration of Greenland Ice-Sheet Flow," *Science* 297 (2002): 218–22.

of Greenland, is the largest drainage channel for interior ice—it alone accounts for about 6.5 percent of Greenland's ice delivery to the sea. Between 1983 and 1997 the ice moved down valley at a speed of three to four miles per year, but then it rapidly increased, reaching about seven to eight miles per year in 2003.[19] An investigation of all the glaciers draining Greenland[20] revealed widespread acceleration of ice loss between 1996 and 2005, first in the south, then steadily moving north. In that decade, Greenland revealed an annual ice deficit—yearly ice losses that exceeded the annual replenishment by snowfall—that more than doubled.

The accelerated ice movement is redefining the phrase "at a glacial pace." The slow flow of ice is similar to the slow creep of Silly Putty down a gentle slope. Both ice and Silly Putty, although solid, yield to the pull of gravity through slow plastic deformation in their interiors. But ice complains when "a glacial pace" is scaled upward—it literally creaks and groans as gravity pulls it downward at a faster pace. Already this disequilibrium is manifesting itself with the appearance of ice quakes, small seismic disturbances created as the ice adjusts internally to the changing velocity of flow, and as it squeezes its way over or past irregularities in the rock surface below.

WEST ANTARCTICA

West Antarctica is capped with an immense mound of ice about a mile high and a thousand miles across. From the top of the mound the ice flows in three directions—to the Ronne Ice Shelf in the Weddell Sea, to the Ross Ice Shelf, and to the Amundsen Sea. Both the Ronne and

19. I. Joughin, W. Abdalati, and M. Fahenstock, "Large Fluctuations in Speed on Greenland's Jakobshavn Isbræ Glacier," *Nature* 432 (2004): 608–10.
20. E. Rignot and P. Kanagaratnam, "Changes in the Velocity Structure of the Greenland Ice Sheet," *Science* 311 (2006): 986–89.

Ross ice shelves, even though losing some ice on their seaward margins, have remained effective buttresses to the ice streams feeding them. But the two principal glaciers carrying most of the ice flow from West Antarctica to the Amundsen Sea are showing dramatic acceleration. The Pine Island and Thwaites glaciers are not typical Alpine glaciers that flow a few tens of miles in valleys a mile or so across. Multiply those dimensions by ten to envision the immense scale of the Pine Island and Thwaites glaciers. These ice streams, each tens of miles wide and up to a mile thick, flow for hundreds of miles from the top of the West Antarctic ice pile to the sea. The scale is difficult to appreciate—imagine the entire Mississippi River floodplain from Memphis to New Orleans filled with ice several thousand feet thick, slipping southward to the Gulf of Mexico at the glacial pace of about a mile each year. These glacial streams drain an area that would contribute about five of the twenty feet of the potential sea-level rise represented in all the ice of West Antarctica. And the discharge from these glaciers has been speeding up.

The ice shelf fed by these glaciers, like shelves elsewhere in the Arctic and Antarctic, is breaking up, allowing the ice from the interior of West Antarctica to drain more quickly to the sea. An understanding of what leads to the breakup of ice shelves, and the subsequent opening of the "floodgates" to upstream ice, is slowly emerging. It is a complex process involving not only meltwater penetration through surface crevasses and moulins, but also thinning and weakening along the bottom of the floating ice sheet by warmer seawater. These new observations are challenging our old understanding about the rate at which glacial ice flows.

Earlier in this chapter I mention that there was an important caveat attached to the IPCC estimate that the likely sea-level rise in the present century would be in the range of ten to thirty inches. In making these estimates, the IPCC considered only the effects of continued thermal expansion of ocean water, the addition of new water by melting and runoff of continental ice, and the return of some groundwater to the sea. Such estimates have proved conservative in the past—actual sea-level rise since 1990 has been at the upper limit of earlier projections by the IPCC. In their 2007 report, the IPCC purposely sidestepped evaluating the accelerating ice loss and the direct deposit of ice into the sea from Greenland, the Antarctic Peninsula, and West Antarctica, which would lead to both larger and faster sea-level rises.

The reason for this sidestep lies in part in the IPCC rules—very new research results, those that have not been around long enough to be digested and evaluated by other researchers, cannot be included in the reports. The IPCC was, however, very much aware of the implications that accelerated ice loss held for rapid sea-level changes, and posted a clear cautionary statement that their estimates "did not include the possibility of significant changes in ice-flow dynamics." But as the ice behavior on the fringes of both Greenland and Antarctica is showing, this possibility may already be contributing to an acceleration of the rise of sea level. Scientists in the field, and at conferences and workshops, have quickly turned to evaluating the changes in ice dynamics, and are revising projections of ice loss upward.

EARTH HISTORY IS replete with examples of different climates and higher sea levels, as evidenced in part by sedimentary rocks of various ages that had been deposited in shallow seas lapping onto the continents. The fossil-bearing sandstones, shales, and limestones that drape the older continental crust give testimony to higher levels of the sea several times during the past six hundred million years. In the Late Cretaceous period, ninety to one hundred million years ago, sea level was five hundred feet higher than today; warm ocean waters washed over interior North America from Alaska to the Gulf of Mexico, and dinosaurs flourished in the surrounding wetlands. And at other times in Earth's history there were extensive glaciations, including the possibility that Earth was at one time frozen over entirely.[21]

Higher seas or extensive ice at different times in the past, however, do not offer as much insight into the more recent climate change as one might hope for because in earlier eras, Earth was very different from Earth today. Continents and ocean basins were in different locations on the globe, and ocean currents that move heat around the globe had distinctly different patterns, because of the constraints placed on circulation by the position of the continents. Earth's modern climate system reflects these fundamental present-day constraints: an ocean at the North Pole, a continent at the South Pole; a Gulf Stream that transports tropical heat to the polar north, and an Antarctic Circumpolar Current that prevents the warming of the polar south.

But if there is disappointingly little guidance to be gleaned from the ancient past, there is much to be learned from the geological record of the more immediate past. Just prior to the beginning of the most recent glaciation, about 120,000 years ago, Greenland had only about half its present ice cover, and sea level stood some 10 to 15 feet higher than today. At the time there were, perhaps, a few hundred thousand people on Earth, mostly still in Africa. Today, upward of 400 million people live on terrain that would be inundated by that large a rise in sea level.

The ice cores extracted from the Greenland ice cap reveal some surprises in the ice from the very bottom of the ice sheet—fossil DNA of ancient trees, plants, and insects that lived in southern Greenland a half-million years ago,[22] later to be obliterated by ice advances during more recent glaciations. This evidence of a forest in an area now covered with thick ice clearly indicates that the ice volume on Greenland has oscillated—sometimes less, sometimes more than in the present day—

21. P. F. Hoffman et al., "A Neoproterozoic Snowball Earth," *Science* 281 (1998): 1342–46.
22. A. de Vernal and C. Hillaire-Marcel, "Natural Variability of Greenland Climate, Vegetation, and Ice Volume During the Past Million Years," *Science* 320 (June 20, 2008): 1622–25.

with obvious implications for increases or drops in sea level. But few of our ancestors lived anywhere close enough to Greenland to notice the comings and goings of the ice, and those living far from ice but closer to the sea were mobile enough to contend with moving shorelines. Nomadic hunter-gatherers made no investments in permanent dwellings.

Earth also experienced a significant warm interval some three million years ago, in the middle of the Pliocene epoch of geological time, and sea level was as much as one hundred feet higher than today. The Atlantic and Gulf coastal plains of the United States were partially flooded, with the shoreline close to one hundred miles inland from its present position. If Augusta, Georgia, and Richmond, Virginia, were Pliocene cities, seaside beach homes would have been only a few minutes away. The Florida peninsula was entirely submerged. The sediments deposited in this shallow sea include shark teeth, the fossils of sea turtles, and pollen, which indicated the proximity of nearby land.

In contrast to the climate systems that governed ice and sea levels much earlier in Earth's history, the mid-Pliocene climate is an excellent analog of today's global climate system—the locations of the continents have changed only slightly over the past three million years. The uplifting of the Isthmus of Panama had just closed the connection between the Atlantic and Pacific oceans, the Drake Passage between South America and Antarctica had been open for several million years, and oceanic currents had achieved a distinctly modern pattern. The microscopic single-cell marine foraminifera in the deep-sea sediments beyond the continental shelf show the periodic oscillation in oxygen isotopes, indicating that the coming and going of modern-style ice ages had begun. These micro-fossils from three million years ago also revealed a surprise—that the glaciations of the Pliocene were dominated by the 41,000-year cycle in the tilt of Earth's axis, unlike the three most recent glaciations paced by the 100,000-year cycle in the orbital ellipticity. The more rapidly oscillating climate of the Pliocene may have been the stimulus for the evolutionary change from the earlier Australopithecines to the modern genus *Homo*. In the mid-Pliocene there was no ice in the Northern Hemisphere, and reduced ice on Antarctica. The greatly elevated mid-Pliocene sea level and the much-diminished ice indicate how much the ice can change and the seas can rise, within the constraints of the modern climate system.

The overarching lesson of the Pliocene is sobering: an ice-free Northern Hemisphere, with no sea ice covering the Arctic Ocean and no ice sheet on Greenland, is a possible condition of the modern climate system. When this happened in the Pliocene the global average temperature was only about four to six Fahrenheit degrees warmer than today. And that may be where we are headed again. If at times in the past ice ruled the world, then in the warm centuries of the future, seawater—ice's playmate on the global hydrological seesaw—will be the formidable adversary of human life on Earth.

CHAPTER 8
CHOICES AMID CHANGE

If you do not change direction, you may end up where you are heading.

—LAO TZU
Chinese philosopher,
sixth century BC

Where indeed are we headed? The ark of humanity seems dangerously adrift in the sea of climate change, with no apparent navigational charts, or even a captain, on board. Can we prevent a world without ice? Can we avoid flooded coastlines? Are there pathways to the future that are less calamitous? These are straightforward questions, but ones that do not yield simple answers.

CONFRONTING UNAVOIDABLE CHANGE

Unfortunately, there is no course of action that will freeze today's status quo and forestall any further changes. Change is under way and is certain to continue because of inertia in both the climate system and the global industrial economy; it is impossible simply to pull the plug and stop these systems in their tracks. They each have momentum analogous to that of an aircraft carrier trying to change course—for several seconds after the helmsman turns the ship's rudder to a new heading, the vessel plows ahead on its old course before slowly beginning to turn. In the global climate system, some of this inertia derives from the greenhouse gases we have already emitted into the atmosphere in times past, but with effects that extend far into the future. Carbon dioxide, the principal greenhouse gas, lingers in the atmosphere for more than a century, as it slowly dissolves into the ocean and is gradually consumed by green plants. These greenhouse gases will continue to warm the atmosphere and oceans even if new emissions could somehow be eliminated.

We have no illusions, however, that eliminating greenhouse gas emissions is within easy reach. Climate scientists have developed projections of how the climate might evolve if greenhouse gas concentrations in the atmosphere could be frozen at their current level, to see what changes might accompany a stabilized greenhouse. In climate science and policy circles, this idealized conceptualization of the future is called a "climate commitment."[1] Even after stabilizing the greenhouse, the inertia of the climate system will continue to drive climate change for several centuries into the future.

1. Other definitions of the "climate commitment" have been tied to stabilization of greenhouse gas *emissions* at some level, as opposed to the more common reference to stabilization of the atmospheric *concentration* level.

So what is it that we have already and unavoidably committed ourselves to? Were we able to immediately stabilize the atmospheric greenhouse at current concentration levels, Earth's atmosphere would still warm by about one Fahrenheit degree by the end of this century. This warming would be followed by continued loss of Arctic sea ice, shrinking of the ice caps in Greenland and West Antarctica, ocean waters warming to greater depths, changes in the geography and intensity of storms and drought, and sea level rising at a rate almost twice that experienced in the twentieth century. Because this climate commitment is the outcome of an unachievable assumption—an immediate and complete stabilization of greenhouse gases in the atmosphere—we must recognize that the projected outcomes constitute an underestimate of the changes that will actually take place. In other words, the consequences that will actually unfold in this century, while still veiled in some uncertainty, will exceed those just mentioned. It is imperative, therefore, to begin planning for changes that are unavoidable, an endeavor broadly termed adaptation.

STRATEGIES OF ADAPTATION

Adapting successfully to a changing climate will require fundamental and sweeping reassessments. We need to ask how a changing climate will affect everything we do, wherever we do it. For example, in agriculture the questions might include:

- What problems and opportunities will a longer growing season present?
- What will be the impact of warmer soil on seed germination?
- What problems will changes in water availability and timing create?
- Will different crops be better suited to the future climate than the current crops?

- Will different pests and weeds replace those present today?
- Will there be a need for more or fewer, or different, fertilizers?

Already researchers are looking to wild relatives of some domestic foodstuffs in search of the genetic makeup that has given these wild plants natural resiliency to drought, heat, and changes in the salinity of water.[2] Other research has focused on building resiliency to extended flooding, with some strains of rice now able to survive up to two weeks underwater, compared to only a three-day survival of earlier varieties. And of course careful water management is an essential in any setting—researchers and farmers are always studying and testing new methods of irrigation.

In the public health arena, practitioners will need to think about how to cope with increased frequencies of extremely hot days and nights in crowded cities, how to safeguard municipal water supplies against bacterial blooms, how to modify sewage systems to cope with more extreme rainfall events, and how to prevent more frequent food contamination in a warmer world.

The list has no end. The transportation infrastructure of the Arctic region, long dependent on hard frozen ground or thick ice, must adapt to a softer, mushier foundation for much or all of the year. Emergency preparedness agencies will need to reshape their responses to more frequent floods, hurricanes, and wildfires, and make plans for refugees from rising seas. Electric utilities will face higher peak demands during the more intense and more frequent heat waves, while at the same time adjusting to different sources of electricity generation. City managers, urban planners, and architects will need to rethink what is required to build or reconstruct climate-resilient and energy-efficient cities, and governments of oceanfront municipalities will face zoning and building code issues along beaches and barrier islands, and infrastructure changes to

2. For a good discussion of agricultural adaptation, see Nathan Russell, "The Agricultural Impact of Global Climate Change," *Geotimes* 52, no. 4 (April 2007): 30–34.

their harbors. The private sector will find opportunities to provide new materials, products, and technologies.

The marine fishing industry will have to anticipate where the fish will be found as the temperature structure of the oceans changes. Freighters on the Great Lakes, in response to lower lake levels resulting from increased evaporation, will have to lighten their loads to access shallower harbors, and modify navigation to avoid hazards that once lay safely below the surface. The insurance industry is already facing a revision of its risk and rate tables as floods become more frequent, wildfires more widespread, hurricanes more intense, the storm season longer, and coastal areas more vulnerable to storm surges as sea level rises. Some insurance carriers have already pulled out of the home insurance market in Florida because climate-related threats—and the consequent cost of claims—are too great. And educational institutions at all levels will bear new responsibilities to prepare students for the demands of a changing world.

The unavoidable future also includes issues that humans have never had to think about in the past. Near the top of the list is open access to the Arctic Ocean. There is a very real possibility that in only a few decades the Arctic Ocean will be free of ice in the summertime, giving people unimpeded access to this vast region for the first time in human history. In 2007, the extent of summer sea ice diminished to the lowest ever recorded since comprehensive synoptic data have been available. In 2008, for the first time in at least a half century and probably much longer, a ring of open water encircled the Arctic—both the Northwest and Northeast passages were open simultaneously.

This physical opening of the Arctic Ocean leads to an opening of important geopolitical issues as well—the claiming of territory, the exploration for mineral and energy resources, and the exploitation of biological resources—issues that were more or less moot when the Arctic was inaccessible. The nations bordering the Arctic already have begun jockeying for position. In 2007, a Russian submersible planted the national flag in the ocean floor at the North Pole, an action reminiscent of when, a half century earlier, the

USS *Nautilus* surfaced at the North Pole and the United States opened a scientific research station at the bitterly cold and windswept South Pole. While largely symbolic, occupancy of the pole with its central 360-degree range of vision is a geopolitical statement of control.

In 2008, the U.S. Geological Survey released a study of oil and gas potential in the Arctic Ocean. This study indicates possible Arctic oil reserves equal to three years of current global consumption, and perhaps a decade of natural gas reserves. The latter amount equals the vast land-based natural gas reserves in the Russian Arctic. Fortunately, from the point of view of potential conflicts, the USGS said that most of the reserves are located in areas on the continental shelf where national sovereignty is well established, principally in offshore Alaska and offshore Russia. The biggest exception is the Lomonosov Ridge, a relatively high submarine topographic feature that extends from Asia toward Arctic Canada and Greenland, bisecting the Arctic Ocean. Russia, Greenland, and Canada all have asserted that it is part of their continental shelf, and are seeking exclusive mineral rights under the provisions of the United Nations Convention on the Law of the Sea.

It is not just energy resources that can generate tension in the Arctic. At a time when the historical fisheries of the world have been severely depleted, the biological resources of the Arctic are becoming increasingly attractive and accessible. The countries and peoples bordering the Arctic Ocean have a long-standing dependence on food from the sea, and will welcome access to new resources. But so will many others; Japan, Korea, and China, nations with large fishing fleets, are sure to cast their eyes on—and their nets in—the Arctic Ocean. The European Union has already had discussions about how to oversee the development and exploitation of the Arctic, as well as to provide protection of the environment and the indigenous people.[3] NATO representatives meeting in Iceland in 2009 discussed security challenges that might develop as the Arctic opens, and

3. See Adele Airoldi, "The European Union and the Arctic" (Copenhagen: Nordic Council of Ministers, 2008), ANP 2008:729.

Canada has already revealed plans to build a deepwater port and a military training center in the high Arctic. Both the Pentagon and the United States polar research community have called for building a much larger icebreaker fleet to enable greater access to and better control of U.S. polar waters.[4] Not to be outflanked, Russia, too, has announced plans to deploy military forces to protect its national interests in the Arctic.

To many people adaptation means planning for the long-term future, but there are some places where adaptation is already a necessity. About halfway south along the western coast of Greenland is the small town of Ilulissat, home to some five thousand residents and an equal number of sled dogs. Ilulissat, known also by its Danish name of Jakobshavn, is the third largest town on Greenland. Tom Henry, a reporter for the *Toledo Blade*, wanted to help his Ohio readers appreciate the consequences of climate change in the Arctic, to make sure they knew that this was as much a story about people as it was about polar bears. In 2008, Henry persuaded his publisher to send him to Ilulissat to get a look at a place where both climate change and adaptation are current realities. In the Inuit language, *Ilulissat* means "icebergs," appropriately so because the town sits near the mouth of the fjord that hosts the Jakobshavn Glacier—Greenland's most prolific ice stream. This glacier alone accounts for more than 6 percent of the ice loss from Greenland's interior ice cap, an amount that has doubled in only the past decade.

Once in Ilulissat, in many conversations around town, Henry learned of both changes and adaptations. Most Inuits in the surrounding areas travel by dogsled over the flat-surface sea ice, because much of the rugged inland topography makes for difficult, if not impassable, sledding. Unfortunately, the diminishing sea ice has effectively isolated residents of outlying settlements for much of the year. Fishermen docked in Ilulissat are finding the halibut more elusive and the catch more expensive, as the warming ocean water

---

4. Anita Jones, "An Icy Partnership," *Science* 317 (September 14, 2007): 1469; Andrew C. Revkin, "Experts Urge U.S. to Increase Icebreaker Fleet in Arctic Waters," *New York Times*, August 17, 2008.

has driven the fish to greater, and cooler, depths. But the tourist industry is booming as visitors come to see massive icebergs break from the glaciers, and to watch whales that now cruise the area, consuming large quantities of fish species unknown around Ilulissat prior to the warming of the water.

NAVIGATING AN UNCERTAIN FUTURE

Given that it will be effectively impossible to hold greenhouse gases in the atmosphere at current levels, what can we expect from more realistic scenarios? The answer to this question, like many other attempts to illuminate the future, are burdened by many uncertainties—uncertainties in our knowledge of how the climate system works in all its complexity, in our ability to transcribe what we do know about climate into detailed computer models, and in how we humans, major players in today's changing climate, will respond to the challenge it presents.

The nonscientific public often has been impressed with the successes of science: the highly precise and accurate predictions of solar and lunar eclipses centuries into the future, of the comings and goings of Halley's comet, and of the transits of Venus across the face of the Sun. Science has enabled the realization of ballistic missile intercepts far out in space, the launch and management of satellites that have revolutionized mobile communications, and the pinpoint navigation of the *Phoenix* Lander to Mars. Given these achievements, it's not surprising that many people have high confidence that scientists will lead us smoothly and accurately into the future, with few surprises.

But in reality it is not an easy task to forecast the future, particularly the future of a system as complex as Earth's climate. Such a large natural system is generally far more complicated than the relatively simple physics governing the orbits of celestial bodies and the trajectories of spacecraft. There is also a big difference between predicting the future of inanimate systems such as planetary orbits and that of a system in which humans play a very important role. When human behavior is part of the equation, the uncertainty of the outcome escalates substantially. So, when the IPCC scientists declare that global average temperature at the end of the twenty-first century will likely exceed the temperature at the beginning of the century by 3.2 to 7.2 Fahrenheit degrees, the range expresses the uncertainty not only about the climate science, but also about how the people and governments of the world will address the challenges of a warming climate.

The latter uncertainty, associated with the human responses to climate change, is sometimes called behavioral or social uncertainty. This uncertainty about the future arises first from an uncertainty about how Earth's human population will grow over the next several decades. Over the past ten thousand years the global population has doubled almost eleven times. The most recent doubling, from four to eight billion people, began in 1975 and will probably be achieved around 2020. At the beginning of this century demographers projected the population at mid-century would reach somewhere between eight and eleven billion people, most likely around nine billion. But what a huge uncertainty that is—a difference of three billion people between the lower and upper estimates. That uncertainty is equal to the entire population of the globe in 1960. Because energy consumption is directly related to the number of people on Earth, clearly this demographic unknown casts considerable uncertainty into the mid- and end-of-century estimates of temperature increases associated with the growing use of energy.

But it is not just the number of people who will populate Earth that is uncertain—there is also uncertainty about how much energy each person will consume in the future. All peoples of the world aspire to an easier and better life, an aspiration that is usually possible via access to affordable energy. The entire history of per capita energy consumption nearly everywhere has been one of ever-increasing growth—rapid in modern industrial societies, slower in remote rural regions. The future trend, with very few exceptions, is a continuing acceleration in energy consumption. How rapidly per capita energy consumption will grow throughout the present century is another source of uncertainty in the projections of climate change.

How the energy will be generated is yet another source of uncertainty. Will it continue to come from coal, petroleum, and natural gas—the carbon-rich fossil fuels—or will it come from renewable carbon-free energy sources such as wind, solar photovoltaics, geothermal, and nuclear? A transition from fossil to renewable energy sources must traverse a minefield of politics and economics, and of regional and industrial-sector special interests. In 2009 the United States made a political transition from an administration that for almost a decade did little to move away from reliance on carbon-based energy, to an administration that appears willing to embrace non-carbon energy alternatives.

THE MOST RECENT Intergovernmental Panel on Climate Change report comprises three volumes, each about the size of the New York City telephone directory. More than two thirds of the pages are devoted to aspects of how the future might unfold. The sections of the report dealing with the consequences, impacts, and mitigation of climate change, and adaptation to it, all examine various demographic, political, economic, and technological pathways into the future. The IPCC does not have a mandate to recommend policy—its responsibility is only to make projections about the future under a variety of social scenarios.

The scenarios vary widely. At one end of the spectrum is a twenty-first century that includes high population growth, continued reliance on carbon-based energy, slow economic development and technological change, and little international integration of regional economies. This scenario, sometimes called the business-as-usual pathway, depicts an accelerating rate of greenhouse gas emissions, CO_2 levels in the atmosphere that reach between three and four times the preindustrial level by the end of this century, and a range of severe climate consequences.

Stephen Schneider, a climate scientist at Stanford University, calls this the "worst-case scenario."[5]

At the other end of the spectrum is a scenario with a global population that peaks in mid-century at around nine billion people, and then declines slowly in the second half of the century. It depicts a rapid introduction of conservation measures and new energy-efficient technologies, widespread development of non-carbon-based energy sources, and strong growth in an integrated, globalized, and increasingly service- and information-based economy. This scenario would result in a much lower rate of greenhouse gas emissions that peak before mid-century and fall thereafter. Levels of CO_2 in the atmosphere would remain below double the preindustrial level, and would produce less severe but far from trivial consequences. Several other scenarios fall between these two bookends, with variations in one or another of the demographic, economic, or technological components. The IPCC placed no probabilities on which of the scenarios, if any, will represent the way the twenty-first century will actually unfold.

MODELS OF FUTURE CLIMATE

For any given emissions scenario, climate scientists can project how the temperature, ice distribution, sea level, precipitation patterns, and many other aspects of climate will evolve in the future by simulating the entire climate system on a powerful computer. The radiant energy from the Sun; the aerosol and dust loading of the atmosphere; the current distribution of ice and vegetation around the globe; the equations of heat and mass transfer in the atmosphere, oceans, soil, and rocks; and how they interact and make exchanges with one another and with terrestrial life forms—all are represented in hundreds of thousands of lines of computer code. Given a particular input emissions scenario, what

5. Stephen H. Schneider, "The Worst-Case Scenario," *Nature* 458 (2009): 1104–5.

comes out of a simulation is the climate of the future, as envisioned by the scientists who write the computer code. In other words, the climate projections are the output of a computer "model" of the global climate system constructed by a group of climate scientists.

Many scientific groups around the world have developed such models, some with great complexity, others with less. Each model expresses the best judgment of the science team creating it—judgment about how to simplify complexity without sacrificing accuracy, about how to represent computationally awkward equations more simply, about how much regional detail to strive for without unduly increasing the time it takes the computer to do the calculations. These different judgment calls lead to different projections for the climate.

Which of these different model projections, describing conditions a century or more into the future, will prove to predict the evolution of the climate with the greatest accuracy? We cannot know, because the future has yet to unfold. To fully appreciate why projections of the future are always expressed in terms of a range of outcomes, we must assess the uncertainty not only associated with the different social scenarios, but also that arises from differences among the climate models. Blunt, definitive statements that declare "this is the way it's going to be," without any mention of uncertainties or probabilities, should always be viewed with suspicion.

Because models represent the real world incompletely and imperfectly, and yield predictions that are embedded in uncertainty, we must always evaluate the predictions with careful scrutiny. Following computer models with unwavering rigidity can lead to cliffs of disaster. We need only think of the highly touted financial models that failed to foresee the partial collapse of the securities and capital markets in 1998, and the near-total paralysis and failure of these markets again in 2008. Both collapses had a common theme—most banks failed to recognize the fragility of their loan portfolios. This myopia, however, can be traced to the economic models that underestimated the risks associated with all kinds of loans, and lured banks, hedge fund managers, and investors big and small onto thin economic ice. That ice eventually and dramatically gave way, sending the entire global economy to depths not experienced for many decades.

George E. P. Box, a well-known statistician at the University of Wisconsin, once stated bluntly: "All models are wrong. Some are useful." To extract utility from models, one must always be skeptical of their structure, and strive to recognize their likely limitations. Emanuel Derman and Paul Wilmott, two experienced builders and users of economic and financial models, assert that the most important questions about any model are, What does it ignore and How wrong is it likely to be?[6]

These caveats about financial models apply to environmental and climate models, too. Orrin Pilkey, a coastal geologist at Duke University, says that computer models of shoreline retreat in the face of rising sea level[7] don't even approach reality. I, too, have some reservations about the way many climate models handle the heat exchange between the soil and atmosphere. And my University of Michigan colleague Joyce Penner, also a contributing author to the IPCC assessment reports, offers a well-informed opinion that most global climate models fail to represent fully the complex effects of atmospheric aerosols on the radiative forcing of the climate system. But neither Penner nor I categorically reject climate models because of their imperfections—we both recognize their utility, indeed their necessity, in spite of their current limitations.

In a wide variety of fields, computer models are extremely versatile quantitative tools used with great success. Meteorologists now employ computer models to give us very reliable forecasts of the weather up to a week in advance, and to plot the likely trajectories of hurricanes as they approach populated areas. Geologists use sophisticated numerical models to map the likely subsurface pathways of plumes of contaminated groundwater, and petroleum engineers employ computer models

6. Emanuel Derman and Paul Wilmott, "Perfect Models, Imperfect World," *BusinessWeek*, January 12, 2009, 59–60.

7. Orrin H. Pilkey and Linda Pilkey-Jarvis, *Useless Arithmetic: Why Environmental Scientists Can't Predict the Future* (New York: Columbia University Press, 2007).

to determine the optimal exploitation of oil and gas resources deep beneath the Earth's surface. Even the future reliability of stockpiled nuclear weapons is determined in part with complex computer models.

The hard reality is that computer models are the only effective tools we have to explore quantitatively the large range of possible scenarios about how future climate change will unfold. We should not be dismayed by the imperfections of the models, or distracted by the uncertainties surrounding the results. Just because scientists, demographers, economists, and policymakers don't know everything, that doesn't mean that they know nothing. They clearly do not operate in a state of complete ignorance; to the contrary, they have substantial knowledge in their fields of expertise. With an appropriate dose of humility that openly acknowledges uncertainties in a straightforward way, and encourages repeated probing for weaknesses in the structure of the models, climate models will continue to be very useful and instructive.

REDUCING UNCERTAINTY

All too frequently one hears skeptics or politicians present uncertainty as an excuse to avoid making important policy decisions. It is important to recognize, though, that postponing important decisions because of uncertainty is actually just an implicit endorsement of the status quo, and often an excuse for maintaining it. It is a fundamental bulwark of the policy known as business-as-usual. Waiting for climate change uncertainties to disappear is not a feasible option, because much of the uncertainty, particularly the social uncertainty, will never go away. We cannot know with certainty what the population will be fifty years from now, nor can we know with certainty what technological innovations will emerge.

Can we expect that future research will yield a better understanding of how the climate system works? Can we anticipate bigger and faster computers that will require fewer compromises in the climate model computing codes? Yes, we certainly will see improvements over time, but they are unlikely to lead to significantly improved model projections that would make the wait worthwhile. Improved climate models that might narrow the range of policy options will be of little help if the improvements come only after the policy opportunity is no longer an option. Uncomfortable as it may be, important policy decisions about how to mitigate and adapt to climate change must be made in the face of considerable uncertainty about the future.

In my 2003 book *Uncertain Science . . . Uncertain World*, I write about how uncertainty both permeates and motivates science, and how it subtly influences people's everyday activities as well.[8] Whether we realize it or not, uncertainty is something we live with and adjust to all the time. Robert Lempert and his colleagues at the RAND Corporation, in a book with the intriguing title *Shaping the Next One Hundred Years: New Methods for Quantitative Long-Term Policy Analysis*,[9] expand these concepts to identify bedrock principles for developing sound long-term policies in the face of deep uncertainty. They reframe the question "What will the long-term future bring?" into a different question: "How can we choose actions today that will be consistent with our long-term interests?" In other words, they provide guidelines that help decision-makers, faced with deep uncertainties, to make sound policy without having answers to every important question.

Because there is deep uncertainty about the future, Lempert and his RAND colleagues argue that we should not try to predict the long-term future with precision, because too many surprises lie beyond the horizon. Winston Churchill captured this perspective when he said, "It is a mistake to try to look too far ahead. The chain of destiny can be grasped

---

8. Henry N. Pollack, *Uncertain Science... Uncertain World* (Cambridge, UK: Cambridge University Press, 2003).
9. Robert J. Lempert, Steven W. Popper, and Steven C. Bankes, *Shaping the Next One Hundred Years: New Methods for Quantitative Long-Term Policy Analysis* (Santa Monica, Calif.: RAND Corporation, 2003).

only one link at a time." We need to explore a wide range of scenarios about how the future might unfold, and to seek strategies that do well in many different scenarios. Finally, we must monitor the impacts of policy actions, and the changing conditions in which the policies are being implemented—and make mid-course corrections as necessary. We learn a lot about how complex systems work by watching how they behave. When the system behavior deviates from a desired pathway, it is time for a mid-course correction to realign the system behavior with our goals. This flexibility is called adaptive management, and it will be critical if the world is to confront climate change effectively.

TIME IS RUNNING SHORT

In 2009 the concentration of carbon dioxide in the atmosphere reached 390 ppm, and was increasing by 2 to 3 ppm each year. The IPCC's assessments of the impacts of higher temperatures due to increasing levels of CO_2 and other greenhouse gases indicate that serious problems in freshwater availability, ecosystem disruption, food production, coastal erosion, and public health—already emergent today—will be very apparent when the level of atmospheric CO_2 reaches 450 ppm. One does not need higher mathematics to recognize that at the current rate of emission—the business-as-usual scenario—CO_2 will reach that level before mid-century, and will continue climbing to even higher levels. The clock is ticking, even as we debate the best course of action.

If we are to have a chance of averting the worst of the consequences of climate change and ice loss, policymakers must make major decisions soon, even without answers to many important questions. Serious reductions in greenhouse gas emissions, described earlier in the more aggressive alternative to the business-as-usual scenario, must take place over the next few decades. Why? Because the lifetime of carbon dioxide in the atmosphere is long, and a few decades of delay will impose centuries of consequences. After the United States squandered most of this century's first decade with a business-as-usual climate policy, there is no time to waste in implementing new energy and climate policies that include serious emission reductions. Such a proactive step is called mitigation.

MITIGATION OPPORTUNITIES

The shadow of an uncertain future, possibly one with extraordinary changes that have severe consequences, provides a motivation for a rapid reduction in and eventual elimination of the human causes of climate change. The principal focus of mitigation is to slow and then reverse the loading of the atmosphere with anthropogenic greenhouse gases. The mechanisms of mitigation are many—some make use of existing technology and are available immediately; others require development of new technologies and will come online later.

Conservation and Efficiency

At the very top of the list of mitigation options are energy conservation and efficiency measures in transportation, manufacturing, household appliances, and buildings. Benjamin Franklin famously said that "a penny saved is a penny earned," and that concept applies to energy consumption as well—a kilowatt-hour saved is a kilowatt-hour that need not be produced, and a gallon of gasoline not used represents dollars that stay in a driver's pocket. The cheapest energy is always the energy that one does not use.

According to researchers at the Lawrence Livermore National Laboratory, more than half of all the energy produced in the United States is wasted.[10] Two thirds of the energy used to generate and distribute electricity is lost before it ever reaches a home to light a bulb or heat a stove.

10. *New York Times*, April 6, 2008. Data from the Lawrence Livermore National Laboratories, Livermore, California.

The personal transportation sector—cars and light trucks—wastes more than 70 percent of the energy contained in gasoline, and the American manufacturers have been notoriously slow to improve fuel efficiency. Imported vehicles have captured an increasing share of sales in the United States for almost half a century, and now account for more than half of the American market. To be sure, the causes of declining market share for American automobile manufacturers go beyond just excessive fuel consumption. But it is fair to argue that the U.S. auto companies' long resistance to higher fuel economy standards hastened the decline of recent years. A doubling of automobile fuel efficiency of American cars is already possible utilizing existing hybrid technology. Even a tripling could be achieved by reducing the weight of vehicles through use of strong, lightweight composite materials. Today, in addition to the driver, most vehicles move at least a ton of steel down the highway. In essence, most of the fuel these vehicles consume goes toward moving themselves, and only incidentally their occupants.

If Americans, indeed people everywhere, were to drive fewer miles each year, they would accrue substantial fuel savings and emissions reductions. Less driving could be achieved in part with greatly expanded high-quality public transportation. Many cities in the United States have been slow to provide viable alternatives to driving personal vehicles. Where such alternatives exist, however, millions of people take advantage of them on a daily basis. The subway in Washington, D.C., which opened in 1976, has become the second busiest rapid transit system in the United States, trailing only the New York City subway. The success of the Washington Metro, as well as newer, well-used light-rail systems in Dallas and Minneapolis, show that a clean, reliable, frequent, and safe rapid-transit system can be a very attractive alternative for many urban and suburban commuters. Even the older, somewhat dysfunctional system in New York remains the most practical, cost-effective choice for millions of daily riders.

If the trend toward suburban housing with long commutes to work could be reversed, more fuel savings and emissions reductions could be achieved. What might promote such a reversal? A revitalization of attractive, affordable housing in city centers. The enduring success of New York as a vital city is in no small part because people live, buy their groceries, do their shopping, and go to school and work in neighborhoods throughout the city—a majority of them without even owning a car.

Fully 40 percent of America's energy consumption is associated with the buildings in which they live and work. Efficiency and conservation measures in the heating and air-conditioning of buildings offer potentially large energy savings. Furnaces, air-conditioning units, and many household appliances are available today that operate well above 90 percent efficiency, in contrast to older units that struggled to reach 50 percent. And upgrading home and building windows and insulation to keep more of winter's cold and summer's heat outside is a low-tech improvement with a rapid payback.

Carbon-Free Energy

Energy sources that do not produce greenhouse gases are of course attractive mitigation options. The carbon-based fossil fuels—coal, oil, and natural gas—are in effect stored solar energy from ages past. All are derived from ancient life forms composed in part of carbon, energized by sunshine, and sequestered underground for millions of years. It should be no surprise that direct utilization of modern sunshine is a principal hope for carbon-free energy.

SUNSHINE

Solar radiation has long been used for direct heating of living spaces and domestic water, but it can also be collected at an industrial scale to produce steam to drive electrical generators. Additionally, solar radiation can be converted to electricity directly by photovoltaic devices, better known simply as solar cells. These devices already provide electrical power for myriad small applications—hand calculators, cell phones and portable radios, sailboats, road signs, remote scientific instruments, and much more. At rooftop scale, solar cells can provide a nontrivial fraction of domestic electricity, even where half the days are cloudy. Improving the efficiency of solar cells is an important and promising research area—today's solar cells convert only about 20 percent of the incoming solar energy into electricity, leaving lots of room for improvement.

WIND

Uneven solar heating on a planetary scale creates differences in atmospheric pressure. The atmosphere responds by pushing air from high-pressure areas to places with lower pressure—a motion we call wind. In places where the wind is strong and steady, there is great potential for generating abundant electricity. Long used in windmills and water pumps, the ubiquitous wind has in recent years fostered development and deployment of modern wind turbines in large "wind farms." Denmark produces about 20 percent of its electricity from wind, and the United States about 2 percent. The technology is improving rapidly, and cost reductions have already made wind price-competitive with carbon-based energy. Wind is the fastest-growing source of new energy-generating capacity worldwide, particularly in Europe and the United States.

FALLING WATER

Hydroelectric power generation at large dams on big rivers, the modern equivalent of hydropower from water wheels, today provides almost 20 percent of the world's electrical energy. But its potential for growth is limited; most of the best locations already have such installations. The tidal movements of ocean water are in a few places driving electrical generators, and researchers are developing prototype devices driven by river currents. Emerging technologies will also soon capture the up-and-down motion of waves along some coastlines to generate electricity.

NUCLEAR ENERGY

When atomic bombs brought World War II to a sudden end, the world witnessed the almost unimaginable energy unleashed from the nucleus of fissionable elements. In weapons, the energy is liberated in explosive fashion, but the process of splitting a nucleus can also be controlled to liberate energy in a slow and steady stream. Worldwide, nuclear energy generates about 14 percent of the global electricity. In the United States, the world's largest producer of nuclear-generated electricity, about 20 percent of the nation's electricity comes from nuclear installations. France generates more than three quarters of its electricity using nuclear energy, but even that large fraction of France's electricity is less than the electricity generated by U.S. nuclear plants. The expansion of nuclear-generating capacity around the world faces several hurdles, including the very high capital costs of construction, a need for large volumes of water to cool the reactor, operational safety concerns, and the complex challenges of securely storing waste that will remain dangerously radioactive for thousands of years.

EARTH'S HEAT

Just ten feet below the surface, Earth barely feels the seasonal oscillation of the surface temperature, from winter to summer and back again. The temperature at that depth sits stably at the year-round average of the surface temperature. In winter the underground temperature is higher than at the surface, and in summer it is lower. That characteristic, a subsurface temperature that does not change seasonally, is the basis for geothermal home heating and cooling systems. In winter, heat is extracted from the warmer soil to heat the house, and in summer heat is removed from the house and returned to the soil. Essentially the system is a two-way heat pump that exchanges heat with the surrounding soil via water circulated through a closed loop of buried piping.

Another type of geothermal energy is the heat contained in very hot rocks near volcanic magma, in places only a few hundred feet beneath the surface. This extreme subterranean heat can produce both hot water and live steam that can be captured to heat buildings or generate electricity. The Geysers Geothermal Area, seventy-five miles north of San Francisco, provides much of the electricity for coastal California north of the Golden Gate Bridge. In Iceland, the island nation located in the middle of the Atlantic Ocean just south of the Arctic Circle, geothermal waters warm most of the houses and buildings. Even without nearby magma, the temperature of rocks everywhere rises with increasing depth beneath the surface. These warm rocks are also viewed as a potential source of thermal energy to heat water for industrial and domestic use.

BIOMASS

For millennia, people have burned wood to provide heat and light, and later to generate steam to power machinery. But trees are only one of many plants that can yield energy through combustion. It may seem counter-intuitive that biomass offers possibilities for mitigation of greenhouse gas emissions—after all, plants have much the same carbon-based composition as the ancient plants that comprise coal. But the production of energy from biomass does in principle provide emissions mitigation, because it just recycles carbon dioxide—extracting it from the atmosphere as the tree or plant grows, and returning it to the atmosphere when burned as a fuel—with zero net increase in atmospheric CO_2. By contrast, burning of ancient coal sends fossil carbon into today's atmosphere, and is thus a net addition of CO_2. But not all biomass has the same energy content, and not all processes to extract that energy are equally efficient. For example, ethanol produced from corn—after taking into account all the energy needed to grow the corn and produce the fuel—is barely a break-even operation. Corn-based ethanol has another downside: diversion of cropland and a primary edible grain into energy production, thereby exacerbating the daily reality of hunger for tens of millions of people around the world. Fortunately, other non-food vegetation, including some hardy weeds and even green algae growing in bodies of water, hold considerable promise as biomass fuel sources.

Capturing Carbon

With an enormous amount of coal available around the world, many ask if there could not be a way to continue using that abundant resource, but somehow prevent the combustion products, including CO_2, from escaping into the atmosphere. Can we not somehow capture the CO_2 and contain it harmlessly somewhere? Trapping carbon and storing it safely is the dream of the so-called "clean coal" campaign.

Storing carbon may be the easier half of this mitigation strategy. Storage, or sequestration, takes two forms: biological storage and geological storage. Plants store carbon as they grow. Hardy forests, with trees that live many decades or even centuries, are in effect warehouses holding significant carbon. About 20 percent of the CO_2 growth in the atmosphere is attributed to worldwide cutting of forests; consequently, slowing or reversing deforestation could take a substantial bite out of the steady growth of atmospheric CO_2. Carbon can also be stored directly in the soil, with attendant benefits to both the soil and the atmosphere.

Geologic sequestration involves pumping of CO_2 underground into rock formations with sufficient tiny pore spaces to accommodate large volumes of the greenhouse gas. Natural gas companies already use underground storage to adjust supply to meet seasonal demand. Gas produced in the summer is stored underground, to be available during the peak demand of winter. This storage strategy has been well tested—nature has stored natural gas underground for millions of years. Several field tests are now under way in rock formations beneath the North Sea and at several sites in the United States and Canada, to test the practicability of large-scale CO_2 storage. Deep ocean basins have also been considered as repositories, because liquefied CO_2 is denser than seawater at the pressures encountered in that environment. But issues of long-term stability and of changes in ocean chemistry have yet to be resolved.

In order to store CO_2 it must first be captured. The technology to pull CO_2 from smokestacks where it is generated, or directly from the atmosphere where it accumulates, is still in its infancy. A joint industry-government project launched in the United States in 2003 to demonstrate the feasibility of "clean-coal" electrical generation, complete with carbon capture and storage, continues but has not yet reached a proof-of-concept stage. Small pilot projects using a variety of technologies to capture carbon show promise, but the difficulties in full-scale development and deployment remain.

Slowing Population Growth

The extraordinary growth of the human population in the twentieth century, along with each person's ever-growing appetite for more energy, has made humans the greatest agent of change on Earth. Obviously, one approach to reducing the demand for energy would be to slow the rate of growth of Earth's population. The number of people on Earth of course plays a big part in the human footprint on the planet, as noted in chapter 6. But discussions of population levels have never been formal agenda items at any international conference addressing climate change. There simply are too many political and religious pressures that have kept population planning off the table for discussion or negotiation.

WILL THESE VARIOUS mitigation strategies be fast enough and comprehensive enough to rein in greenhouse gas emissions over the next two to three decades? All mitigation strategies have strengths and shortcomings, proponents and detractors. If the long debates about whether to require higher fuel efficiency in automobiles or where to store nuclear waste are any indication, the urgency of confronting climate change may be blunted by political pushing and pulling that in the end may deliver too little, too late.

Debates about which of the mitigation strategies offers the best chance of reducing emissions miss the point: we need them all. If we hope to avert the harsh consequences of climate change, we need every horse in the stable pulling together, and as hard and as fast as possible. Ironically, the severe global economic instability that began in 2008 may promote a greater willingness to take bold steps that may dramatically reshape America's energy infrastructure and industrial economy. On the other hand, the economic distress might instead serve as an excuse for further inaction on climate change. That would be a tragedy of historical proportions, because there truly is no more time to waste.

ACCELERATIONS

Because the future is burdened with uncertainty, we must be particularly observant of the way the real world is behaving, and always be assessing how well the model projections compare with reality, how well the assumptions implicit in the model continue to be valid. Consider the simplest type of model projection of some quantity X into the future—one in which the rate at which X changes remains steady, and so the cumulative change in X is just proportional to the passage of time. In technical terms, this is called a linear extrapolation, because a graph of the changes in X over time will be a straight line, upward sloping if X is growing, and downward sloping if X is decreasing.

How likely is it that the processes affecting X will continue to change at the same rate? There is no rule of nature that requires such a linear relationship to continue forever. Just as a small tree branch will bend a little when a boy steps out on it, and will bend a little more when his girlfriend joins him, everyone knows that there is a limit to the loading, beyond which the branch no longer bends—it snaps. Slow, incremental change may lead to greater and more rapid change as some limit is approached or crossed.

Scientists look for evidence of changes in the rate at which things are happening—either slowing down or speeding up. Such changes are called decelerations and accelerations. Changes in rates are often the first hint that a system is no longer behaving as it did before, and may be about to change abruptly and dramatically. For example, we should be very alert to increasing rates of atmospheric and oceanic warming and ice loss.

Year-to-year observations of Earth's vital signs are providing much evidence of accelerating changes. The average rate of warming of Earth's atmosphere over the past 150 years has been almost 0.1 Fahrenheit degree per decade, but the rate of warming over only the past century is 60 percent higher than over the 150-year period. And over more recent intervals, the acceleration is even greater—during the past 50 years, Earth warmed 2.8 times faster than the 150-year rate, and over the past 25 years, almost 4 times faster.

The use of energy in the United States has also accelerated throughout the twentieth century. For every unit of energy consumed by a person at the beginning of the century, by 1960 the per capita consumption was four times greater, and by the end of the century it was almost seven times greater. Because the growth of population over the century is already taken into account in per capita statistics, this acceleration in energy consumption is wholly attributable to changes in standard of living and lifestyle—driving more and in bigger cars, eating more food transported over longer distances, and living in bigger houses with more electrical appliances.

On the population front, it took more than 10,000 years for the population to reach 1 billion people. But it took only 130 years more for the population to reach 2 billion, and another 32, 15, 13, and 12 years for it to reach 3, 4, 5, and 6 billion. Between 1980 and 1990, the growth in Earth's population averaged more than 80 million each year, the highest growth rate in all of human history. But since that decade, there is a hint of deceleration. The annual population increments have begun to decline slightly—in 2004, population grew by about 75 million—and the United Nations is projecting a continuing decline in the growth rate, to roughly 30 million additional people per year by mid-century.[11]

Not surprisingly, the growth in atmospheric CO_2 reflects both the population and energy consumption trends. The Keeling curve that shows the growth of atmospheric carbon dioxide over the past five decades (shown on page 184) also shows acceleration in the growth rate. When Keeling first started his measurements, the rate of growth was just under 1 ppm per year, but today the CO_2 level is increasing at more than 2 ppm per year, a doubling of the rate of growth in just a half century. The rate at which sea level is rising is also accelerating. In the fifty-two years from 1961 through 2003, sea level rose almost four inches, one third of which occurred in the last decade alone. Sea-level changes, of course, are related to ice loss from the continents and warming of the deep oceans, so an increase in the rate at which the seas are rising implies faster rates of ice loss and ocean warming in recent decades.

The extent and thickness of Arctic sea ice are both diminishing at ever-faster rates, and although the loss of sea ice does not directly raise sea level (sea ice is already floating), there are important indirect effects that do lead to rising seas. Less Arctic sea ice in the summer means that more ocean water is exposed to absorb solar radiation, and the refreezing of this warmer water will take place later in the fall. And newly frozen sea ice, thinner than sea ice that survived the summer breakup, will also break up earlier the next summer. Earlier breakup and delayed refreezing results in a longer warming season for the open ocean water. This warming eventually mixes into the deeper ocean and leads to sea-level rise through thermal expansion.

And as discussed in chapter 7, glacial ice from Greenland, the Antarctic Peninsula, and West Antarctica is being delivered to the sea at accelerating rates. The ice shelves that impeded ice loss from the continents have been disintegrating rapidly in the last decade, allowing land-based ice to spill into the sea and raise sea level. This speedup in

11. United Nations Population Division, *The World at Six Billion*, 1999.

the flow of ice to the sea came as a surprise to glaciologists[12] and led the Intergovernmental Panel on Climate Change to caution in its 2007 report that its projections of future sea level did not take into consideration the possibility of rapid changes in glacial ice dynamics. Because the 2007 IPCC estimate of twenty-first-century sea-level rise, less than three feet, did not include any contributions due to accelerated delivery of land ice to the sea, that estimate clearly must be recognized as a rock-bottom estimate, which may well be exceeded.

Only a few years have elapsed since the IPCC report appeared, and it may already be outdated. In a special 2009 assessment[13] of possible sea-level changes in the twenty-first century, the U.S. Climate Change Science Program pointed out that since 1990, the global rate of ice loss has been more than double the rate observed from 1961 to 1990. If ice spillage to the sea continues throughout this century at the rate observed in its first decade, enough ice will enter the oceans to raise sea level three feet. And to that rise must be added the thermal expansion of the seawater as the oceans continue to warm—an effect that will raise sea level at least as much as the new ice does. Both effects together will raise sea level some six feet in the present century, compared to a rise of less than a foot in the twentieth century.

Are there hints of other unpleasant surprises in the near future?

TIPPING POINTS AND CLIMATE SURPRISES

Just as the international financial system surprised the world with a major collapse in 2008, the global climate system, with its human component,

12. R. B. Alley, M. Fahenstock, and I. Joughin, "Understanding Glacier Flow in Changing Times," *Science* 322 (2008): 1061.
13. U.S. Climate Change Science Program Synthesis and Assessment Report 3.4, "Abrupt Climate Change, Chapter 2: Rapid Changes in Glaciers and Ice Sheets and Their Impacts on Sea Levels," 2009. http://www.climatescience.gov/Library/sap/sap3-4/final-report.

is equally capable of serious surprises. Lurking in the shadows of climate change is the possibility that the accelerations we now observe in the climate system are portends of approaching tipping points.

Tipping points represent changes in a system that occur when the system passes from one mode of behavior to another, sometimes imperceptibly, sometimes suddenly. A simple analogy is the process of paying off a home mortgage. Each monthly mortgage payment comprises both interest and principal. In the early years of the mortgage, the payoff of the loan principal is painfully slow and annoyingly incremental, as most of the monthly payment goes to paying the interest on the loan. In a typical thirty-year home mortgage, a homeowner, after ten years of payments, has paid off only 10 percent of the loan. After twenty-one years of payments, the monthly check is finally split evenly between interest and principal, a tipping point that typically passes without recognition or acknowledgment. But beyond that tipping point the reduction of the unpaid balance accelerates, and as the mortgage approaches payoff, there is a rapid erosion of the remaining unpaid loan. At the end there is another tipping point, impossible not to notice—a very abrupt transition to a new state in the homeowner's personal finances, when there is no mortgage payment to make at all.

In the climate system there are several possible tipping points: major realignments of oceanic and atmospheric circulation, rapid releases of greenhouse gases now trapped in permafrost and in the ice that exists at shallow depths beneath the ocean floor, and sudden changes in sea level. All these possibilities are related to changes in Earth's ice.

What role does ice have in taking the climate across a tipping point? The average temperature of a planet's surface depends directly on the amount of incoming solar energy absorbed by the surface. But not all the solar radiation delivered to Earth is absorbed—some is reflected back to space. Snow and ice are both highly reflective substances, and so the fraction of Earth's surface covered by snow and ice is a big determinant of Earth's average surface temperature. The more radiation that is reflected away, the less energy remains to warm the planet. Currently, Earth reflects about 30 percent of the arriving solar radiation back into space.

When the amount of snow and ice cover changes over time, so does the balance between reflection and absorption of solar energy. As ice increases on Earth, more solar energy is returned to space and less is absorbed, thus lowering the surface temperature. More ice promotes a cooler planet, and a cooler planet encourages the accumulation of even more ice. This interaction is called a positive feedback, and leads to an ever-faster acceleration of climate change. Diminishing ice cover also drives a similar feedback, but in the other direction: as Earth becomes darker and less reflective, more solar radiation is absorbed, the planetary surface grows warmer, and a warmer planet leads to even less ice cover and a further acceleration in warming.

How do the ice-climate feedbacks lead to tipping points in the climate? As we have just seen, the loss of sea ice in the Arctic Ocean is allowing much more solar radiation to be absorbed in the Arctic summer, causing a warming of the Arctic Ocean. But the principal circulation pattern of the Atlantic Ocean is strongly dependent on dense Arctic seawater sinking to make room for the warm surface current—the Gulf Stream—traveling northward from the tropics. Increased summertime warming of Arctic seawater, however, makes the water more buoyant and less inclined to sink. As this Arctic warming continues, eventually the Arctic seawater will not sink. When that happens, there will be no room in the Arctic for warm water coming from the south—and the Gulf Stream will weaken or shut down. The consequence? A deep and enduring chill would descend over Western Europe.

It may seem counterintuitive that warming of the Arctic could lead to a cooling of Western Europe, but Europe occupies a latitude band roughly similar to central Canada and central Asia, regions with much harsher climates. Europe is warmer than those colder regions because it draws heat from the warm waters of the Gulf Stream. A slowdown or shutdown of the Gulf Stream would again place Western Europe into the refrigerator, as during the Younger Dryas episode 12,700 years ago, when the Gulf Stream was interrupted and European temperatures dropped by about ten Fahrenheit degrees. What starts as a local phenomenon in the seawater of the high Arctic quickly affects the circulation of the entire Atlantic Ocean and the climate of Europe.

Another feedback with the potential to similarly alter Atlantic currents relates to the melting of the Arctic permafrost. Melting of this permanently frozen ground across vast expanses of Canada, Alaska, and Siberia is already under way. The melting provides more freshwater to flow into the Arctic Ocean via the Mackenzie River, which drains much of western Canada, and the Lena, Yenisei, and Ob rivers, which drain northern Asia. Because freshwater is also more buoyant than saltwater, the Arctic Ocean, already more buoyant because it is warming, is being made even more buoyant by the increased freshwater input from the melting permafrost. The warming and freshening reinforce each other to impede the sinking of Arctic Ocean water, and thereby slow the Gulf Stream.

How likely is this major change in oceanic circulation? The IPCC simulations show several scenarios projecting a 25 percent slowdown in circulation by the end of the century, but none that project a complete collapse. But even with a slower oceanic transport of heat to the high latitudes, the increased greenhouse warming of the atmosphere will likely compensate, and spare Western Europe from cooling, at least for a while.

The melting of the permafrost has the potential for yet another major impact on the climate system—the release of large volumes of the greenhouse gas methane to the atmosphere. Strengthening the greenhouse effect would lead to more atmospheric warming, which in turn would lead to continued reduction of the permafrost, more methane release, and thus an even hotter greenhouse. Another vast source of methane lies trapped in a form of ice present in sediments at shallow depths beneath the ocean floor. But if the ocean water warmed sufficiently to release the methane trapped in the ice, the methane would quickly bubble to the surface and lead to an even stronger greenhouse.

The existence of the methane-bearing seafloor deposits is well established, and the geological record hints that these deposits were destabilized around 55 million years ago,[14] producing a stronger atmospheric greenhouse by an amount roughly equivalent to all the carbon that has been released to the modern atmosphere since the beginning of the industrial revolution. This intensified greenhouse caused Earth's surface temperature to rise some nine to fifteen Fahrenheit degrees—a hot spell that lasted for more than 100,000 years. This event, which geologists call the Paleocene-Eocene Thermal Maximum, was probably the last time Earth was entirely without ice.

What is the likelihood of a sudden methane release occurring in the near future? Methane has been observed bubbling out of the continental shelf into the Arctic Ocean in many places, and out of the permafrost in Siberia as well. But the physical processes by which permafrost and subsea ice can be destabilized are generally slow, and thus large and abrupt releases seem unlikely. Submarine landslides can expose and rapidly destabilize the methane-bearing formations, but the geographic extent of landslides is usually small. Most simulations of methane liberation from the seafloor show it will not even begin until the ocean bottom water warms by a few degrees, and then the release will likely extend over tens of thousands of years. Fortunately, that is a slow process, unlikely to accelerate.

How stable is Greenland's ice? Several computer simulations of future melting there show a temperature threshold beyond which the Greenland ice sheet passes a point of no return. Once that threshold is crossed—probably in this century—the melting will have such inertia that the Greenland Ice Sheet will likely disappear completely, in a relentless meltdown extending over several hundred years. Beyond that tipping point, the surface of Greenland will inexorably show more rock and less ice, and shorelines will relocate as sea level rises. And by then there will be nothing humans can do to stop it. To return to the nautical

14. Quirin Schiermeier, "Gas Leak!" *Nature* 423 (June 12, 2003): 681–82.

analogy, it would be like watching two ships at sea approaching each other, belatedly realizing they were on a collision course. Even though both may frantically try to steer a new course to avert colliding, they have passed the point when course corrections can take hold in time—their inertia will drive them on to the collision.

Those simulations assume that melting is the only way that Greenland will lose ice. They do not take into account the possibility of bulk ice loss to the sea prior to melting—an omission that is becoming increasingly questioned in the face of the current acceleration in the delivery of bulk ice to the ocean. The observed acceleration of ice loss from Greenland, the Antarctic Peninsula, and West Antarctica is putting to rest the idea that in order to raise sea level, land ice must first melt and the meltwater then flow to the sea. We are seeing the early stages of glacial ice sliding directly into the ocean at a rate much faster than it is being replenished by snowfall inland, an observation that strongly suggests another acceleration may be soon apparent—a further increase in the rate that sea level is rising. An ominous message comes from coral reefs that were living 120,000 years ago, during the very final stages of the warm interglacial interval that existed prior to the most recent ice age. These reefs experienced an eight-foot rise of sea level in only fifty years, most likely due to extremely rapid sloughing of ice into the sea.[15]

Greenland is undergoing both increased melting over its surface and a speedup of ice delivery to the sea. In Antarctica, save for the Antarctic Peninsula, the climate is generally much colder than in the Arctic, and surface melting rarely occurs. But much of the ice along the perimeter of East and West Antarctica sits directly on the ocean floor; with only modest thinning, some of this grounded ice could begin to float, lifting off the seafloor and admitting water beneath the ice. Glaciologists have long known that what happens at the base of a glacier affects the

15. P. Blanchon et al., "Rapid Sea-Level Rise and Reef Back-stepping at the Close of the Last Interglacial Highstand," *Nature* 458 (2009): 881–85.

speed at which it flows over the land, but they are only now learning how dramatically the loss of ice can be affected by the incursion of seawater beneath. Effectively the seawater erodes the ice from below, just as warm air can melt it from above. Such an attack from below would almost assuredly lead to an acceleration of ice loss from the interior and faster rises in sea level.

The implications of a rapid acceleration in ice loss from Greenland or Antarctica are profound. The ice in each region alone could contribute more than twenty feet of global sea-level rise; together they could raise sea levels over forty feet, enough to submerge a three-story building. This incursion of seawater would flood coastal cities around the world, and transform New York into New Venice. Only 120,000 years ago, in the warm interval before the last ice age, Greenland lost half its ice and sea level rose ten to fifteen feet. Some three million years ago, during the Pliocene warm interval, sea level was one hundred feet higher. The much smaller and more mobile human progenitors living near the sea at those times adapted simply by moving to higher ground. There were no permanent structures, and certainly no cities anywhere.

But the world today is very different. Millions of people now live at the ocean's edge, in many of the world's largest cities, from Shanghai to New York to Buenos Aires. These modern urbanites might be able to accommodate and adapt to a twenty- to forty-foot rise in sea level over a thousand years, but they would find it nearly impossible to deal effectively with such a rise in only a century, let alone in a few decades. The differences between these scenarios are stark: on the one hand a perhaps orderly adaptation of physical and social infrastructure to an evolving global problem, versus a rapid physical and social disintegration leading to chaos the world over. And the difficulties will not be confined to the shoreline—cities farther inland at higher elevations will be spared the direct inundation, but not the flood of refugees and the resulting social stress arising from the dislocation of hundreds of millions of people fleeing the encroachment of the sea.

CLIMATE ENGINEERING

Confronted with rapid changes in so many of Earth's vital signs, and well aware of the possibilities of impending tipping points and climate surprises, there are some who think that the mitigation measures now on the table will inevitably be a day late and a dollar short. They believe that only planetary-scale "climate engineering" offers hope of averting the worst of climate change consequences. Who are these people, and what do they propose?

These would-be climate engineers are not kooks from the fringes. The list includes respected scientists such as Michael MacCracken, the former director of the U.S. Global Change Research Program; Tom Wigley, a senior atmospheric scientist at the National Center for Atmospheric Research; Ken Caldeira, an atmospheric scientist at the Carnegie Institution, Gregory Benford, a physicist at the University of California–Irvine; and Paul Crutzen, co-recipient of the 1995 Nobel Prize in chemistry—all smart, serious people very worried about Earth's changing climate. Crutzen's Nobel-winning research illuminated the complex chemistry of how the man-made chlorofluorocarbons led to the development of the ozone hole over Antarctica. It was likely an important step in the evolution of his thinking about how humans have become the dominant agents of change on Earth—and his embrace of the term *Anthropocene* to describe the rapid ascendancy of humans in geological history.

So what kinds of large-scale "climate engineering" projects do these scientists have in mind? Their proposals fall into two broad categories: the first addresses ways to prevent sunshine from reaching Earth, and the second focuses on ways to speed up the processes by which Earth stores carbon. The several sunscreen schemes generally try to increase Earth's reflectivity—by sending many millions of tiny mirrors into high orbit, or by spraying seawater into the atmosphere to nucleate more cloud cover, or by shooting sulfate aerosols into the atmosphere to simulate the Sun-blocking effects of volcanic eruptions. There has even been the tongue-in-cheek suggestion that we should no longer try to curtail industrial pollution of the atmosphere, the logic being that dirty air and smog allow less sunshine to reach the Earth's surface. Critics of these sunscreen approaches to climatic amelioration point out that these schemes do nothing to mitigate other environmental consequences of rising carbon dioxide levels, particularly in the oceans, where the trend toward acidity continues, and the marine biosphere is being stressed.

One idea for enhancing carbon storage is large-scale fertilization of the oceans with iron. Theoretically this would stimulate the growth of phytoplankton—organisms that pull CO_2 out of the atmosphere for nourishment, thereby diminishing the greenhouse effect and cooling the planet. Small-scale experiments with iron fertilization do show some enhanced phytoplankton growth, but much of the additional biomass soon decays and returns the captured CO_2 to the atmosphere. A second storage proposal would add calcium to the oceans to react with dissolved CO_2, to promote the formation of limestone. In effect, this would amount to a speeding up of the geological weathering process that Earth's natural thermostat uses to supply calcium to the sea. After precipitating limestone, the oceans then pull CO_2 out of the atmosphere to replace the CO_2 used in making the limestone, thereby reducing the strength of the greenhouse and slowing climate change.

But there is considerable and justified concern about potential unintended consequences of such global engineering. To cast a medical analogy, these proposals would be classified as experimental drugs, with unproven efficacy and perhaps unanticipated side effects. We have to be very careful that the cure is not worse than the disease. More than normal caution should be attached to these large-scale engineering schemes, with which we have little relevant experience. Recall that the problem these ideas might ameliorate, the change in Earth's climate brought by the consumption of fossil fuels, is itself an unanticipated side effect of an inadvertent geochemical experiment—the removal of long-sequestered underground carbon to burn for energy, and allowing the resulting oxidized carbon to take up residence in the atmosphere and oceans.

IS IT THE FATE of the world to lose its ice? If an ice-free world comes to pass, future generations will gaze over vast areas of the planetary surface that have not seen the light of day or felt the warmth of sunshine for thousands or even millions of years. They will see the drab, gray rock beneath Greenland and Antarctica slowly rebound from deep topographical depressions imposed by the heavy load of glacial ice. But these same generations will also watch low-lying areas of the continents being flooded by the sea—areas that have not been submerged beneath the ocean since the Pliocene, or the Paleocene, or the Cretaceous, or perhaps ever. These generations will be forced to confront the political and social challenges of the millions of climate refugees displaced inland.

Some observers see climate change as the greatest challenge the human race has ever faced. They ask if humans, indeed the entire planet, will survive. I do not worry about planet Earth surviving—it has survived many challenges over its long history, including significant impacts by wayward meteorites, asteroids, and comets. I have little doubt that Earth will be making its annual journey around the Sun for millions if not billions of years into the future. So planet Earth itself should not be described as fragile. Rather, it is the great diversity of life that has evolved on Earth—a web that supports human civilization—that is at risk. As the great tectonic plates slowly moved continents, reconfigured oceans, and uplifted mountains, opportunities for new life emerged and the vulnerabilities of some existing life forms were exposed. Some life flourishes amid the stimulus of great geological changes, while other forms falter. The almost seven billion people constituting *Homo sapiens* are soon to be tested.

Imagine, as did geologist Don Eicher, all of Earth's history as events compressed into a single calendar year:[16]

> On that scale, the oldest rocks we know date from about [late January]. Living things first appeared in the sea [in February, and continents began to assemble and drift about the globe in early March. All of the major phyla of marine life had evolved by mid-October, and the generation of petroleum followed soon thereafter]. Land plants and animals emerged in late November and the widespread swamps that formed the great coal deposits of the world flourished for about four days in early December. Dinosaurs became dominant in mid-December, but disappeared on the 26th [shortly after the time the Rocky Mountains were first uplifted]. Manlike creatures appeared sometime during the evening of December 31st, and the most recent continental ice sheets began to recede from [the Great Lakes area] and from northern Europe about 1 minute and 15 seconds before midnight on the 31st. Rome ruled the western world for five seconds from 11:59:45 to 11:59:50. Columbus [reached the New World] three seconds before midnight, and the science of geology emerged with the writing of James Hutton just slightly more than one second before the end of our eventful year of years.

In this compressed perspective of our planet's long 4.56-billion-year history, we humans show up only in the early evening of December 31. In our extremely brief time on Earth we can look at our achievements with some pride—but we must also look at our missteps with trepidation. What might arguably be called our greatest success—the creation and distribution of almost seven billion of us around the world—is also the root of our greatest challenge. It is not altogether clear that the human

16. I first saw the idea of compressing all of geologic time into a single year in *Geologic Time*, by Don L. Eicher (Englewood Cliffs, N.J.: Prentice Hall, 1968). Here I have modified Eicher's compression a little, but it remains true to his "year of years."

race has the vision, determination, or discipline to meet the self-created challenges of climate change and rising seas, or to make the choices that will preserve the social structure that we call civilization. Will later intelligent life forms judge our brief time on Earth, the Anthropocene, only as an excessive New Year's Eve party, which ended at midnight? Or will we humans enter a new era, perhaps with a hangover, but also with a sober resolve to find a sustainable path to the future? The choice is ours.

PATHWAYS TO THE FUTURE

Peoples around the world have of course confronted challenges and made their choices in the past. As colonists in the New World, Americans decided to shape their destiny by breaking away from Great Britain to become a new nation. Only a decade later, the citizens of France rejected centuries of monarchy and chose a democratic path to the future. In the twentieth century, the people of Russia experienced a cataclysmic end to feudalism, and then embarked upon another seven-decade social experiment with communism that ultimately failed as badly as their feudal monarchy did. Today China is engaged in a great social transformation, to a new and not yet fully defined future. All of these changes followed a long and slow accumulation of seeds of instability that eventually crossed thresholds and unleashed rapid change. These all were challenges among people, within the human social structure—none was a confrontation between humans and the natural world.

Now we are immersed in another disruption of the human social fabric—the global financial crisis. It has caught the attention of the world like a whack to the head with a two-by-four. Just as in the earlier political revolutions, the financial instability that became abruptly apparent to everyone in late 2008 was preceded by decades of slow, largely unnoticed erosion that allowed the global economy to pass a threshold into painful collapse. The warnings from a few economists of an impending burst of the financial bubble went unheeded—as are the warnings of today's climate scientists about ice loss and rising seas. As the financial crisis continues to unfold, it is exposing the many unsustainable and risky practices that slowly undermined the world of global finance. But in forcing a review of what led to financial instability, the crisis also is providing an opportunity to develop a clearer vision, one that may enable us to see the interconnectedness between the financial world and the natural world.

History may describe the collapse of the global financial system in 2008 as the meltdown of the twenty-first century. But history will also record that another and ultimately far more significant meltdown—the loss of ice the world over—was already under way. The money spent rebuilding the global financial system, as great a sum as it seems to be, will pale in comparison with the cost of adapting to the warmer world with higher seas and destabilizing human dislocations that will come, if effective and timely mitigation measures are not implemented.

Stephen Schneider describes two types of policy mistakes we can make in confronting climate change. The first, which in the world of risk analysis is called a Type A mistake, would be to spend a lot of treasure to address serious climate change, only to find that as the twenty-first century unfolds, climate change turns out to be much more benign than it first appeared. The second, a Type B mistake, would be to adopt a wait-and-see policy, and then discover too late that climate change consequences are every bit as severe or even worse than predicted, thus requiring big expenditures for adaptation and amelioration. Both will turn out to be expensive mistakes, but the second type, leading to widespread loss of life and property, is much more expensive and socially destabilizing than the first.

The rapidly diminishing cohort of business-as-usual proponents is betting that taking no action is the right action. In what amounts to yet another of their trenches of denial, they assert that mitigation and adaptation measures are cures that we don't need and can't afford. If the consequences of climate change indeed turn out to be benign, they will be proven correct.

However, the IPCC reports and even more recent assessments give very low probabilities to a benign eventuality. By contrast, the proponents of immediate and strong mitigation and adaptation strategies want to make substantial expenditures soon to forestall the worst of the consequences, and they will be proven prescient if climate change hits society very hard.

The U.S. government has for years feared making the first type of mistake, spending money on something we may not need. Only now is it beginning to refocus and think about the brutal economic and social costs of even moderate climate change, and how much greater the consequences will be if we do not address them quickly and aggressively. If indecision continues, however, and mitigation opportunities are missed, we will reach another "silent" tipping point where many more dollars will need to be shifted from mitigation to adaptation, an even more expensive proposition.

Some have called the financial crisis and the climate crisis an unfortunate juxtaposition—they lament that because each is such a large problem, the world cannot possibly afford to address both simultaneously. This is simply the latest version of the old canard that asserts that improvements to the environment will lead to lesser profits and loss of jobs. In truth, the juxtaposition of the financial and climate crises has presented an unusual opportunity to rethink the totality of how we interact with the natural world, and in so doing improve both the economy and the environment. Globalization of the economy and globalization of the environment have not developed independently, nor can breakdowns in each be effectively treated independently.

One often hears that the financial crisis may lead to the rebuilding of long-neglected physical, industrial, educational, health-care, and financial infrastructure, but unless it also stimulates new thinking about how we interact with our natural environment, the rebuilding will be incomplete and ultimately unsuccessful.

Al Gore argues that what many perceive as three separate problems—the financial perils stemming from the United States being the world's largest debtor nation, the security risks rooted in our growing dependence on foreign oil, and the global dangers of a changing climate—are in reality three different facets of a single problem: an inadequate and misguided national energy policy. Treating each as a separate problem will slow progress in solving all of them, whereas recognizing them as manifestations of a single problem can lead us to develop a faster and more comprehensive solution. That is a multidimensional opportunity that we must not allow to pass us by.

LAO TZU, the ancient Chinese philosopher, once warned that in the absence of a change in direction, we will very likely end up where we are headed. That captures the essence of this moment in human history. The only climate policy Americans saw from their national government in the first eight years of the twenty-first century was a stubborn commitment to business-as-usual, a policy that brought the ark of humanity eight years closer to the dangerous shoals emerging from climate change. But it is not too late to steer a new course into the open sea of opportunity. Although the inertia of our ark will surely carry us closer to danger, a sharp change of heading today will steer us away from calamity at mid-century.

Although this challenge is new, history holds instructive lessons about ways people have coped with imminent danger in the past. The American response to the surprise attack on Pearl Harbor in 1941 has relevance to our current challenge of confronting climate change—it demonstrated that once a problem gets our attention, we can muster both the determination and resourcefulness to rapidly confront it. Immediately after Pearl Harbor, the United States entered World War II and quickly transformed a peacetime industrial economy into one completely focused on meeting the challenges of a global war. Domestic manufacture of consumer goods ended abruptly, and within months American industries were turning out airplanes, tanks, jeeps, and ships in astounding numbers. By the end of 1943, just two years after the United States went to war, more airplanes were built at a single factory in Michigan than in all of Japan. That should give us confidence that when people understand the severity of a situation they can refocus sharply and master the challenges they face.

But who is steering the ark of humanity today? In a sense, we all are. Just as billions of people through their individual actions have inadvertently caused Earth's climate to warm, so can we humans reverse this dangerous trend. In truth, however, it will not be easy, and will require us all to do much more than just replacing our old incandescent light bulbs with newer energy-efficient fluorescent bulbs. If we hope to preserve the climate system that sustains us, we must revisit individual decisions about where we live and work, about how much space we require to live comfortably, about energy consumption and conservation in our homes, about our transportation choices, about how frequently we travel, and about how many children we will have. Those are all issues that we can address as individuals, as consumers, and as families.

But we need also to augment individual mitigation efforts with changes that can come only from collective action. We need to amplify our individual voices by joining with others to have larger-scale impacts. And there is no bigger megaphone for our voices than the ballot box at election time. The right to choose the people who will run our governments is the most significant tool we have to turn in a new direction. As individuals, we have little voice in determining how the electricity that comes to our homes is generated. But our collective voice can, through the actions of our government, determine how the energy we use is produced and distributed. Individually we have little control over tax incentives and regulatory controls—that playing field is the domain of government. Only governmental action can landscape that field to end the advantages long held by the coal and petroleum industries and offer incentives for investment in conservation and renewable energy. New government policies could place limits on greenhouse gas emissions and promote employment opportunities in enterprises that enhance rather than compete with the natural environment. Only government has the tools to reshape the regional development and transportation policies that would help us reintegrate into the natural world, and to abandon policies that unconsciously encourage us to live separate from it.

Government policies determine the level of support for scientific research, and for science education in our schools, both important elements in meeting the challenges of climate change. Only government can shape a foreign policy that encourages and promotes international cooperation in addressing global problems, including trade policies that set emissions reductions as a precondition of international commerce. And unless governments are willing to provide more educational opportunities for women, and address the cultural and religious taboos that encumber family planning in many places, little progress will be made in slowing population growth.

Many governments and institutions, however, are not agents of change. Instead governments often act only as custodians of stability, and strive to protect the status quo. That is the very definition of inertia. Governmental and institutional inertia, however, fundamentally derive from personal inertia. If we as individuals do not strive for new directions, our institutions will simply carry us in the direction we are headed—toward dangerous irreversible climate change. Our voices need to be aggregated in many settings—schools, universities, religious congregations, labor halls, civic service organizations, investment clubs, corporate shareholder meetings—anywhere and everywhere we can shape public debate. Government officials everywhere, whether elected or not, must hear that people want and need a new course—because without that message, little will happen.

There is a Native American proverb that says we did not inherit Earth from our ancestors, but have only borrowed it from our children. Will we selfishly repay our children with a degraded planet devoid of ice, with seawater washing over our great coastal cities? Or will we pass on a planet that has been rescued from that fate by its people—the same people who inadvertently initiated climate change, but who also recognized their responsibilities to reverse it before the worst of consequences had drowned their shorelines?

Climate change is an intergenerational problem, centuries in the making, yet many people around the world do not even understand that there is a problem, much less that it is rapidly reaching levels of serious consequence. They do not see the growing momentum of the climate system carrying us to unavoidable consequences, a momentum that without mitigation will make even more severe changes irreversible. People in all walks of life and in all regions of the world need a wakeup call, before rising seas lap at their doorsteps.

The world needs to chart a bold course into a new sea of sustainability. Whether Americans like it or not, the United States must provide a clear compass for the global family of nations, through direct and proactive leadership. While it is true that the problems of climate change are not solely American-made, it is also true that there will be no effective solutions without our full engagement. Much of the world is waiting to see what we do, and we must respond boldly, confidently, and quickly. Winston Churchill described a pessimist as a person who "sees difficulty in every opportunity," and an optimist as one who "sees opportunity in every difficulty." While our journey to the future will surely encounter some turbulent seas, we—like Magellan and Columbus centuries earlier—must never lose sight of the fact that we are sailing out onto a sea of unbounded opportunity. Let us all be Churchillian optimists in recognizing opportunity, and at the same time pragmatic realists in addressing the difficulties we will encounter along the way.

We have our work cut out for us. *Carpamus diem!*—Let us seize the opportunity!

ACKNOWLEDGMENTS

I am grateful to Al Gore for contributing the foreword. I have been acquainted with him for almost two decades—not socially or politically, but in the context of our shared interest in climate science. In 1992, as then-Senator Gore from Tennessee, he held hearings about climate change in his position on the Senate Committee on Commerce, Science, and Transportation. I was invited to testify about the relatively new technique of reconstructing past climate using subsurface temperatures measured in deep boreholes (there is a description of this approach to climate reconstruction in chapter 4). He personally conducted the full day of hearings from start to finish, fully engaged and asking many questions that showed remarkable insight into the intricacies of climate science.

As my research into reconstructing the climate of the past from borehole temperatures matured, a few years later my colleagues and I published a global climate reconstruction using data from four continents: North America and Europe in the Northern Hemisphere, and Australia and Africa in the Southern Hemisphere. This paper[1] appeared

1. H. Pollack, S. Huang, and P.-Y. Shen, "Climate Change Record in Subsurface Temperatures: A Global Perspective," *Science* 282 (October 9, 1998): 279–81.

in *Science*, the prestigious flagship publication of the American Association for the Advancement of Science, in 1998.

The day after publication, I received a phone call from a member of Al Gore's staff, who informed me that the then vice president had read the article and would like me to come to the White House to discuss it with him the next day. My first thought was—which of my professional colleagues was playing a well-orchestrated practical joke, with me as the target? But after a few minutes more of conversation, I realized that the call was no hoax. The next day I was in Washington visiting with Al Gore and Neal Lane, the president's science advisor, and with the staff in the Office of Science and Technology Policy (OSTP).

One new acquaintance from that visit was Dr. Rosina Bierbaum, an ecologist and OSTP's associate director for environment, in whose hands the climate portfolio resided. Little did I know at the time that Rosina would later become a colleague of mine at the University of Michigan, when she was appointed dean of the School of Natural Resources and the Environment in 2001. Since her arrival in Ann Arbor, we have had many opportunities to interact scientifically and on issues of climate policy.

Al Gore's film *An Inconvenient Truth* recorded his famous and oft-presented lecture on climate change—its causes, consequences, and possibilities for remediation. At the invitation of Dean Bierbaum, and prior to the appearance of *An Inconvenient Truth*, Gore presented this lecture in Ann Arbor in 2005. He informed the dean then that he was organizing a program—to become known as The Climate Project—to train volunteers to present his lecture in their communities. He sought a science advisor to help with the training, and Rosina suggested me. Thus, seven years after my visit to the White House, my third engagement with Al Gore began.

The Climate Project (TCP) training began in 2006, on the Gore family farm near Carthage, Tennessee, with an initial group of fifty volunteers from all over the United States and from all walks of life—retirees, students, public officials, businesspeople, entrepreneurs, athletes, housewives and househusbands—whose only common denominator was their desire to build public awareness about the realities of climate change. Since then, in a number of subsequent training sessions in Nashville and elsewhere around the world, Al Gore and the TCP team have trained thousands of volunteers to engage their communities in discussions about climate change. I have participated as one of the teaching faculty at several of these training programs.

As I have come to learn, anything Al Gore does he does with his full energy and enthusiasm. He is a lifelong learner, always reading, inquiring, absorbing, and incorporating new knowledge into his endeavors. In truth, he and I cross paths infrequently—years can go by between one meeting and the next. But from my first meeting with him in the Senate hearings, to the later visit with him in the White House, and now the engagements with him in TCP training, I have seen him give full measure and then some to everything he undertakes. Would that we all accomplish as much as he.

I extend special thanks to my wife, Lana, and son, John, for their thorough critiques and edits of the manuscript at many different stages. Both are excellent writers and unsparing with the red pen. They have made the book more personally engaging and less professorially dry. Both have accompanied me to the Antarctic ice and know well the majesty of that white world.

I am indebted to Victoria Wheatley, staffing coordinator for Abercrombie & Kent, Inc., for giving me so many opportunities to spend time in the ice of the world. Kim Robertson Chater, one of my colleagues on several expeditions to Antarctica, and as capable a Zodiac driver as one can hope for in perilous polar waters, is also an extraordinary artist and created the illustrations for this book. Dale Austin, staff illustrator in the Department of Geological Sciences of the University of Michigan, prepared the maps and graphs.

I am grateful to Jason Smerdon of Columbia University, who critiqued the entire manuscript with his customary thoroughness. I am also grateful to my University of Michigan colleagues Dan Fisher, Ted Moore, Jim Walker, Shaopeng Huang, Josep Pares, Kacey Lohmann, and Bruce Wilkinson, with whom I have had many long discussions about climate. I have incorporated some of their research results into this book. Michael Jackson of the University of Minnesota helped me discover how frequently ice appears in literature, in the oddest of contexts.

Of course, no book sees the light of day without the help of capable professionals in the publishing world. My enthusiastic and talented agent, Gillian MacKenzie, helped me craft the initial concept for this book into a successful book proposal. My editors at Avery—Megan Newman, Rachel Holtzman, Travers Johnson, and Jeff Galas—shaped the manuscript into its final and much improved form.

Finally, I extend a broad umbrella of gratitude to the thousands of glaciologists, oceanographers, biologists, geologists, and climate scientists who have over decades studied the ice and water and air and rocks of Earth to learn the operational secrets of this marvelous planet that we all call home.

INDEX

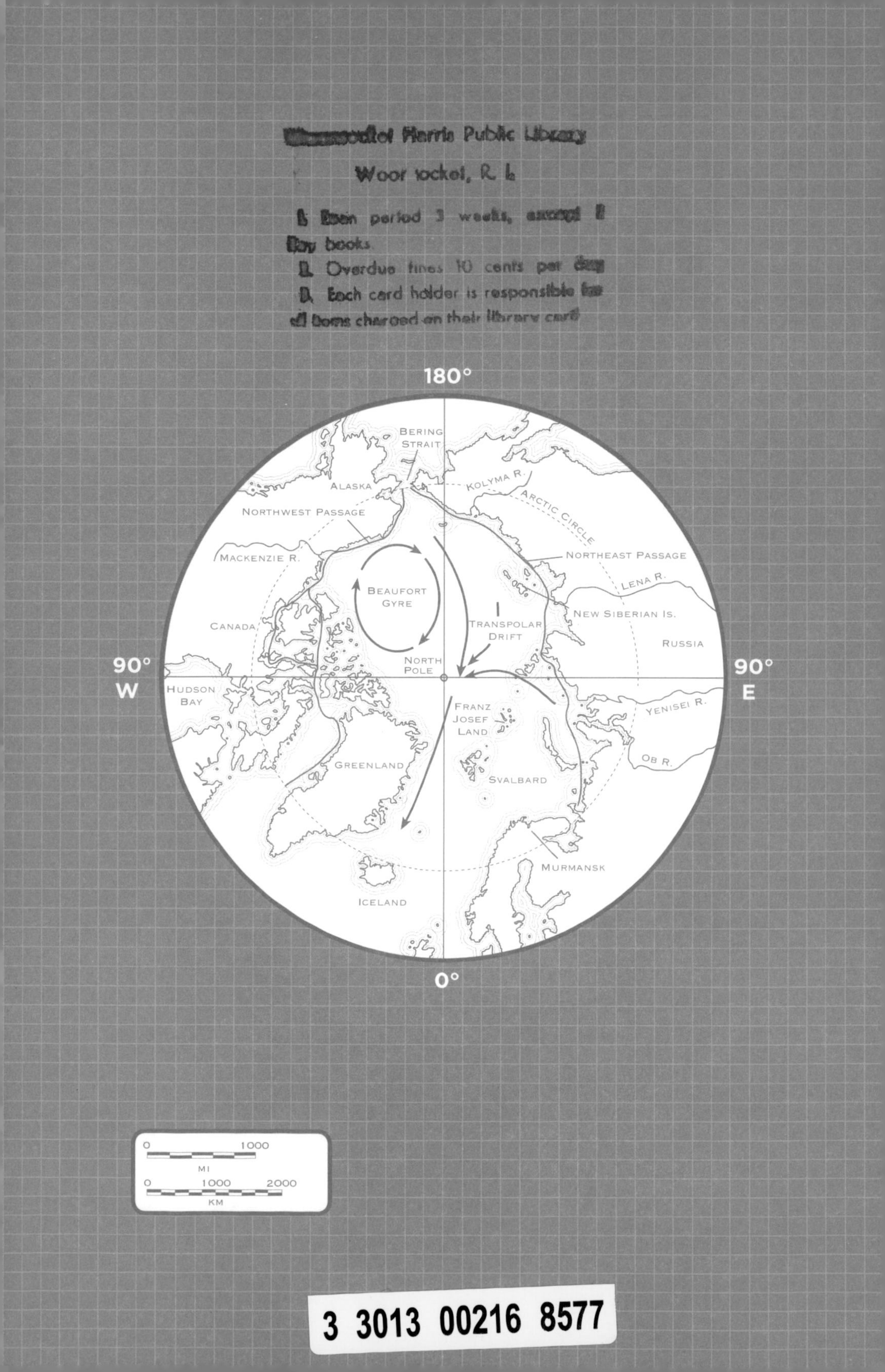
180°
Bering
Strait
Alaska
Kolyma R.
Arctic Circle
Northwest Passage
Northeast Passage
Mackenzie R.
Lena R.
Beaufort
Gyre
New Siberian Is.
Canada
Transpolar
Drift
Russia
90°
W
North
Pole
90°
E
Hudson
Bay
Franz
Josef
Land
Yenisei R.
Ob R.
Greenland
Svalbard
Murmansk
Iceland
0°
0
1000
MI
0
1000
2000
KM